Lecture Notes in Artificial Intelli

T0237808

Edited by J.G. Carbonell and J. Siekmann

Subseries of Lecture Notes in Computer Science

Lutz Maicher   Lars Marius Garshol (Eds.)

Scaling Topic Maps

Third International Conference
on Topic Map Research and Applications, TMRA 2007
Leipzig, Germany, October 11-12, 2007
Revised Selected Papers

 Springer

Series Editors

Jaime G. Carbonell, Carnegie Mellon University, Pittsburgh, PA, USA
Jörg Siekmann, University of Saarland, Saarbrücken, Germany

Volume Editors

Lutz Maicher
Universität Leipzig
Institut für Informatik
Johannisgasse 26, 04103 Leipzig, Germany
E-mail: maicher@informatik.uni-leipzig.de

Lars Marius Garshol
Ontopia AS
Schweigaardsgate 67b
0656 Oslo, Norway
E-mail: larsga@ontopia.net

Library of Congress Control Number: Applied for

CR Subject Classification (1998): I.2, H.4, H.3, J.1, J.3, K.3-4

LNCS Sublibrary: SL 7 – Artificial Intelligence

ISSN   0302-9743
ISBN-10   3-540-70873-1 Springer Berlin Heidelberg New York
ISBN-13   978-3-540-70873-5 Springer Berlin Heidelberg New York

Springer is a part of Springer Science+Business Media

springer.com

© Springer-Verlag Berlin Heidelberg 2008
Printed in Germany

Typesetting: Camera-ready by author, data conversion by Scientific Publishing Services, Chennai, India
Printed on acid-free paper   SPIN: 12443352   06/3180   5 4 3 2 1 0

Preface

The papers in this volume were presented at TMRA 2007, the International Conference on Topic Maps Research and Applications, held October 11–12, 2007, in Leipzig, Germany. TMRA 2007 was the third conference in an annual series of international conferences dedicated to Topic Maps in science and industry.

The motto of TMRA 2007 was "Scaling Topic Maps." Taken literally the motto implies developing Topic Maps approaches that scale to large data and user volumes. This is a very real and useful research problem which is addressed by many of the contributions to the conference. But there is an even broader interpretation of the motto: wide adoption of Topic Maps in academia and industry. This is an equally important problem, and one that the TMRA conference series exists to help solve. And there is a more fanciful view on the motto. To "scale" can also mean to climb, so for the attendees the conference provided a way to "scale the mountain of Topic Maps." In all these ways TMRA 2007 helped to scale Topic Maps.

Stimulated by the success of the previous conferences, the concept of TMRA was retained nearly unchanged. The conference was preceded by tutorials@TMRA, a full day of in-depth tutorials. The main conference schedule was separated into two parallel tracks, providing a rich program for all interests. The Open Space sessions, once more smoothly moderated by Lars Marius Garshol, provided a light and exciting look at work which will most likely be presented at future conferences. And TMRA 2007 was succeeded by a 2-day ISO meeting which emphasized the importance of the conference.

The TMRA 2007 program attracted a very international crowd of around 80 attendees, hosted in the media campus of the Leipzig Media Foundation. The scientific quality of the conference was ensured by the international Program Committee with 35 members. From 44 submissions, 32 were accepted for presentation at the conference. Every submission was carefully reviewed by three members of the Program Committee. After the strict review, 16 full papers, four short papers, the invited keynote, and one invited report from both the poster and open space sessions were accepted to be published in this conference proceedings volume.

We would like to thank all those who contributed to this book for their excellent work and great cooperation. Furthermore, we want to thank all members of the Program Committee, and especially Prof. G. Heyer, for their tireless commitment to making TMRA 2007 a true success. TMRA was organized by the Zentrum für Informations-, Wissens- und Dienstleistungsmanagement at the University of Leipzig. Furthermore, we acknowledge the generous support by all sponsors.

We hope all participants enjoyed a successful conference, made many new contacts, gained from fruitful discussions helping to solve current research problems, and had a pleasant stay in Leipzig. TMRA 2008 will be held October 15–17, 2008 in Leipzig. For detailed information about all TMRA conferences see the website at www.tmra.de.

February 2008

Lutz Maicher
Lars Marius Garshol

Organization

TMRA 2007 was organized by the Zentrum für Informations-, Wissens- und Dienstleistungsmanagement (ZIWD) at the University of Leipzig, Germany.

Program Committee Chairs

Lutz Maicher, University of Leipzig, Germany
Lars Marius Garshol, Ontopia, Norway

Program Committee

Frederic Andres, NII, Japan
Atta Badii, University of Reading, UK
Robert Barta, Bond University, Australia
Michel Biezunski, Infoloom, USA
Dmitry Bogachev, Ontopedia, Canada
Karsten Böhm, FHS Kufstein, Austria
Robert Cerny, Germany
Darina Dicheva, Winston Salem University, USA
Patrick Durusau, University of Illinois at Urbana-Champaign, USA
Eric Freese, LexisNexis, USA
Sung-Kook Han, Won Kwang University, South Korea
Gerhard Heyer, University of Leipzig, Germany
Hiroyuki Kato, NII, Japan
Larry Kerschberg, George Mason University, USA
Peter-Paul Kruijsen, Morpheus Software, The Netherlands
Jaeho Lee, University of Seoul, South Korea
James David Mason, Y-12 National Security Complex, USA
Graham Moore, NetworkedPlanet, UK
Riki Morikawa, U.S. Federal Government, USA
Steven R. Newcomb, Coolheads Consulting, USA
Jan Nowitzky, Deutsche Börse Systems, Germany
Leo Obrst, MITRE, USA
Sam Oh, Sungkyunkwan University, South Korea
Jack Park, SRI International, USA
Rani Pinchuk, Space Applications Services, Belgium
Thomas Schwotzer, FH Brandenburg, Germany
Alexander Sigel, KPMG, Germany
Stefan Smolnik, European Business School, Germany
Eleni Stroulia, University of Alberta, Canada
Volker Stümpflen, MIPS, Germany

Hendrik Thomas, University of Ilmenau, Germany
Markus Ueberall, University of Frankfurt, Germany
Fabio Vitali, University of Bologna, Italy

Sponsoring Organizations

Topic Maps 2008, Oslo, Norway
Bouvet ASA, Oslo, Norway
Networked Planet, Oxford, UK
Media Foundation of the Sparkasse Leipzig, Germany
Ligent, Glenview, USA

Table of Contents

Scaling Topic Maps 1
  Mare Wilhelm Küster and Graham Moore

Applied Topic Maps in Industry and Administration

Convergence of Classical Search and Semantic Technologies – Evidences
from a Practical Case in the Chemical Industry 14
  Stefan Smolnik

A Citizen's Portal for the City of Bergen 25
  Lars Marius Garshol

Visualisation and Representation of Topic Maps

TM*chartis* – A Tool Set for Designing Multiple Problem-Oriented
Visualizations for Topic Maps 36
  Hendrik Thomas, Rike Brecht, Bernd Markscheffel,
  Stephan Bode, and Karsten Spekowius

Open Educational Topic Maps: A Text-Oriented Perspective 41
  Lars Johnsen

Using Topic Maps for Visually Exploring Various Data Sources in a
Web-Based Environment 51
  Eicke Godehardt and Nadeem Bhatti

Collaborative Applications

Topincs Wiki – A Topic Maps Powered Wiki 57
  Robert Cerny

Bookmap – A Topic Maps Based Web Application for Organising
Bookmarks 66
  Tobias Hofmann and Martin Pradella

Standards Related Research

A Theory of Scope 74
  Lars Marius Garshol

Comparing Topic Maps Constraint Specification Languages 86
  Giovani Rubert Librelotto, Renato Preigschadt de Azevedo,
  José Carlos Ramalho, and Pedro Rangel Henriques

Information Integration with Topic Maps

Knowledge-Oriented Middleware Using Topic Maps 98
    Robert Barta

Large Scale Knowledge Representation of Distributed Biomedical
Information 116
    Volker Stümpflen, Thorsten Barnickel, and Karamfilka Nenova

Versioning of Topic Map Templates and Scalability.................... 128
    Markus Ueberall and Oswald Drobnik

Social Software with Topic Maps

Toward a Topic Maps Amanuensis 140
    Jack Park

Cooperative Building of Multiple Points-of-View Topic Maps with
Hypertopic 154
    L'Hédi Zaher, Jean-Pierre Cahier, and Claude Guittard

Metadata Creation in Socio-semantic Tagging Systems: Towards
Holistic Knowledge Creation and Interchange 160
    Roy Lachica and Dino Karabeg

Topic Maps Engines

Ruby Topic Maps 172
    Benjamin Bock

Topic Maps and Dublin Core

Expressing Dublin Core in Topic Maps 186
    Steve Pepper

Mapping between the Dublin Core Abstract Model DCAM and the
TMDM 198
    Lutz Maicher

Information Management with Topic Maps

On Path-Centric Navigation and Search Techniques for Personal
Knowledge Stored in Topic Maps 214
    Jens Heider and Julian Schütte

KAIFIA: Knowledge Assisted Intelligent Framework for Information
Access 226
    *Atta Badii, Chattun Lallah, Oleksandr Kolomiyets, Meng Zhu, and
    Michael Crouch*

Open Space and Poster Sessions

Report from the Open Space and Poster Sessions 237
Lars Marius Garshol and Lutz Maicher

Author Index 253

Scaling Topic Maps

Marc Wilhelm Küster[1] and Graham Moore[2]

[1] University of Applied Sciences Worms, Erenburgerstr. 19, 67549 Worms, Germany
[2] NetworkedPlanet, Oxford Centre for Innovation, Mill Street, Oxford OX2 OJX, UK

Abstract. The *encyclopédie* of Diderot and d'Alembert is unique in building an open discourse model based on maps of knowledge and a wide-ranging coverage of the knowledge of its time. d'Alembert's *Discours préliminaire* prefigured many of the topic map paradigms down to the very metaphor.

However, the *encyclopédie* was still bound by the inevitable restrictions of the print medium. A completely non-literary example, the nascent European eGovernment Resource Network (section 2), shows the need to transcend the encyclopedic model. In many cases we need both distributed registries and repositories. These requirements give rise to prototypical implementations for cross-implementation linked registries (section 3) that gives a hint at how registries can genuinely scale to a distributed world of knowledge.

1 Shaping Knowledge

A servant of Louis XV told me one day that the king his master dined in the Trianon with a small company of his courtiers. The conversation turned to hunting and from there to black powder. 'It is funny', said the Duke of Nivernois, 'that we amuse ourselves every day killing fowl in the park of Versailles, without knowing exactly with what we kill.' Mme de Pompadour added that they were all in the same position. She, similarly, had no idea of what the powder was composed of that she used every day to colour her cheeks.

'It is a pity', said the Duke of La Vallière, 'that Your Majesty has confiscated our encyclopedic dictionaries, that has cost us each a hundred *pistoles*. We would find in it immediately the answers to all of our questions.' The king had three servants fetch a copy of the encyclopedia. With effort each of them carried seven volumes. Everybody threw himself on those volumes, and everybody found there immediately everything he sought.[1]

1.1 The *Encyclopédie de Diderot et d'Alembert*

The *Encyclopédie ou Dictionnaire raisonné des sciences, des arts et des métiers*, famously edited by Denis Diderot and Jean le Rond d'Alembert, was one of the

---

[1] Paraphrased after [1]. While Voltaire's story is entirely fictitious, it demonstrates a lot about the function of the *encyclopédie* in the 18th century.

L. Maicher and L.M. Garshol (Eds.): TMRA 2007, LNAI 4999, pp. 1–13, 2008.

most influential publications of the 18th century and, indeed, of all times. Many historians credit it with its being one of the key sources for preparing the mindset that led to the Revolution and thus to the downfall of the French monarchy (e. g. [2], p. 7).

The *Encyclopédie* was certainly a scientific, political and philosophical endeavour of the first order, but it was also a major economic undertaking. It was one of the most voluminous and expensive publications to have been attempted up to that date in the Western world. 17 volumes in folio of text on 18,000 pages covering no less than 72,000 articles written by more than 140 authors in addition to 11 volumes of images[2] covered important aspects of essentially all of the practical and theoretical knowledge of its time. In this, the *Encyclopédie* represented the thinking and learning of its time to a degree that few other works such as Isidore of Seville's so called *Etymologiarum libri viginti* can claim for their respective periods. It was really a *summa* of enlightenment thought.

1.2   d'Alembert's *Discours préliminaire*

The *Encyclopédie* was much more still, though. Extensive and comprehensive though it was, it was not a mere repository of knowledge. It was just as much and maybe even primarily a means of finding, presenting and linking that knowledge — and it was so very much self-consciously. D'Alembert's preface to the *Encyclopédie*, the *Discours préliminaire* [4], is not only one of the most celebrated forewords in the French tongue, it is also one of the most important studies on the structure, organization and presentation of knowledge ever written — and it has a lot to teach on building and scaling topic maps.

The key question that d'Alembert and his co-editor had to answer when preparing their encyclopedia was both immensely practical and highly philosophical — in what sequence to range the articles. Should they arrange them according to subject matter, and if so how and in what succession of subjects, or should they simply order the articles alphabetically by keyword? Both were at the time well-established options with many, albeit smaller precedents.

Trees of Knowledge: The sequence by subject-matter according to a preestablished tree of knowledge (*arbre philosophique*) was certainly regarded as the more scholarly approach and had dominated both antiquity and the middle ages.

However, which tree of knowledge? Philosophy had seen a great many such ontologies, starting with, at least, the Platonic school. The Tree of Porphyry, named after the third century AD neoplatonic philosopher, was a sophisticated tree of differences popularly taught well into the 19th century, but it was a methodology more than a concrete tree. On a completely different level Isidor had structured his *Etymologiae* according to quite different criteria that deeply influenced medieval teaching.

The tree of knowledge had been the subject of serious study amongst others in the English Renaissance, Lord Bacon's work on the subject being perhaps

---

[2] For the numbers cf. [3].

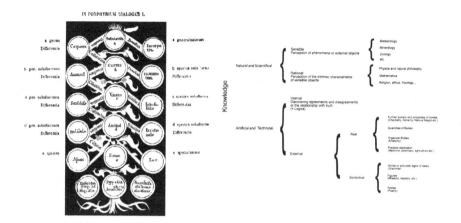

Fig. 1. The Tree of Porphyry (Source: [5]) and Chamber's Tree of Knowledge (Source: [6], p. 532 after [7], p. ii)

the best known, but by no means the only example. The English encyclopaedist Ephraim Chambers [7] would in 1728 retake the thrust and preface his own *Cyclopedia*, the direct predecessor of the *Encyclopédie*, with a tree of knowledge (cf. fig. 1).

However, Chambers was too farsighted a man to actually believe in his tree. Classification for him in contrast to his predecessors was a useful tool, but still "a thing merely artificial, and the work of imagination" ([7], p. ii). To make his point he added a second, quite different tree that for him was just as valid.

A Map of Maps: And Chambers chose to arrange his encyclopedia alphabetically, a decision that d'Alembert emulated. For both it was a conscious protest against one specific knowledge system imposing itself on their respective repositories of knowledge. Any one ontology must violate the multifasceted reality by pressing it into one predefined scheme — "all the lesser parts, one might say, all the parts whatsoever, must be, in some measure, swallowed up in the whole", whereas the alphabet was seen as "perhaps the only way wherein the whole circle [...] of knowledge can well be delivered" ([7], p. II, quoted after [6], p. 533).

D'Alembert echoed this sentiment. While beginning with a detailed exposition of the relationship between direct experience (*nos connoissances en directes*) and theoretic knowledge (*nos connoissances en réfléchies*) and leading on to an overview over the different disciplines, he understands clearly that it is not generally valid. Interestingly, the metaphor he uses to explain his position is precisely the one so well known to all readers of these proceedings — that of a world map of knowledge:

> In the historic order of the mind's progress, it is impossible not to tackle the sciences other than in sequence. This is not the case with the

encyclopedic order of our knowledge. This consists in so to speak placing the philosopher above this vast labyrinth at a very high point where he can perceive simultaneously all the sciences and the principal trades. It is a type of a world map that can show the most important countries, their positions, their mutual interdependencies, the straight line between one country to another one — a way that is often cut by a thousand obstacles. Those obstacles could only been shown in special, highly detailed maps, however. These highly detailed maps are the different articles of the *Encyclopédie* and the tree of knowledge is their world map.

However, just as on the general maps of the globe we live on the different objects are more or less close to each other, depending on where the point of reference is placed by the geographer who draws the map, the tree of knowledge depends on the respective point of view. It is possible to think of as many different systems of the human knowledge as maps of different projects — and each of these systems could even have a particular advantage.[3]

Just as any geographical map any map of knowledge is for d'Alembert a more or less detailed personal view on the underlying territory. It is not entirely arbitrary, though. Just as a map can use any of a number of different projections and scales and thus necessarily reflects personal choices, it is still constrained by the terrain in question and the consistency of its approach and can be valued against both.

For d'Alembert the map is not a direct map of the territory. It is a map of maps, for every article of the encyclopedia in turn represents a more detailed map of its corner of the world of knowledge — a map that shows many more of the specificities, obscurities and internal contrasts of an individual area, but a map nonetheless.

A map with a twist, though. The "real" knowledge out there is not necessarily written and thus difficult to point to. In fact, its nature is by no means trivially understood for d'Alembert, as most[4] learning for him is essentially cumulated, condensed and systematized experience. Scholarly tradition up to then had regularly disparaged practical experience, especially that of the lower classes. Not so d'Alembert and Diderot who recognized that much of the knowledge of their time was in the heads and hands of artisans and as such hardly ever reified. The authors actually went out to the various shops and spoke to the masters, sketched their instruments and machines, jotted down the tricks of the trade. For many of the traditional crafts of the 18th century, from metal foundry to watch-making, the "map" with its detailed descriptions and many excellent images is thus our primary source. For us it has become the territory.

---

[3] [4], p. 109, my translation: The term 'philosopher' in the 18th century was a much broader term than today, in many aspects closer to today's concept of an intellectual than that of a philosopher in our current understanding.

[4] The main exception to this rule is revealed religion (*religion revelée*) which he explicitly includes in his circle of knowledge ("Rien ne nous est donc plus nécessaire qu' une religion révélée qui nous instruise sur tant de divers objets", [4], p. 95).

The Alphabetic Option: So, what is in this context then the role of the alphabet? It is the great randomizer, the tool that breaks up traditional contexts and splits the knowledge into small, seemingly indipendent units held together by many links (*renvois*) between them and the overaching trees of knowledge as their maps. Both together constitute new contexts, partially prefigured by the authors, but concreted only in the mind of the reader.

In this the alphabetic option is constitutive for the *encyclopédie* as it alone allows Diderot and d'Alembert to realize their vision of a printed hypertext, virtually structured by topic maps and finally realized as concrete texts only through the intricate interaction between authors and readers. It is this structured openness that makes the *encyclopédie* both so strikingly modern and such a big conceptual progress over many of its successors to this very day.

1.3 Panckoucke's *Encyclopédie méthodique*

In fact, the direct legal heir to the *Encyclopédie*, Charles-Joseph Panckoucke's *Encyclopédie méthodique* (1783-1832 in 206 volumes), would shut again the cognitive process and replace it with the closed intellectual framework of the rising bourgeoisie, symbolized by philosophers such as Auguste Comte.

> In a period of a few decades we see that [...] the open and the closed discourse clash — not by chance coinciding once with the revolt of the bourgeoisie and once with its final victory ([6], p. 589, my translation)

Panckoucke and Comte replaced the multiplicity of coexisting ontologies with their one true tree of knowledge — a mindset that, astonishingly, persists into much of today's semantic web philosophy where ontologies all too often seen as fixed ("best") ways of ordering the world (or at least a given discipline).

1.4 Beyond the Encyclopedia

If the *Encyclopédie* of Diderot and d'Alembert was a great advance over much that came, it still was limited by the inevitable constraints of print technology and part of what McLuhan was to call the "Gutenberg Galaxy" [8] — even though it pushed its boundaries to the extreme.

Firstly, the encyclopedia was simultaneously a repository of knowledge and a registry, both part of a single enormous endeavour. As such, all of its aspects were tightly coupled and often planned years in advance. While the articles were as a rule elaborated by many authors across France, all were centrally controlled by the main editors.

Secondly, even if both could be decoupled — and companion volumes such as [9] suggest that they partially could — it would still remain a single registry superimposed on a specific, constant body of knowledge that, once published, was essentially frozen for decades to come due to the prohibitive cost of reissuing updates.

Thirdly, the links of the *Encyclopédie* were necessarily internal only, and in their selection and placement again controlled by the work's authors and editors. The links opened the discourse, but primarily within the *oeuvre* itself — references to external publications were not core to the encyclopedia's concept.

In all of these three points we are now in a position to step to transcend the encyclopedia conceptually. We can (at least to a large degree) decouple our work on repositories and registries, with topic maps being an obvious technical choice for the latter.

We can, however, go further still and distribute both registries and repositories across countries and institutions, transforming central registries and repositories into virtual entities. Those entities are composed dynamically out of existing registries and repositories while still maintaining uniform interfaces to the outside world. What is more, we can also superimpose many different systems of hyper-references on that body of knowledge, again modelled as topic maps.

The authors will demonstrate both the need for distributed registries and repositories using a completely non-literary example, the nascent European eGovernment Resource Network (section 2), and show prototypical implementations for linked registries (section 3).

2   A Use Case: The eGovernment Resource Network

When looking at the relationship between data repositories and linked registries, it is useful to do so in the light of a concrete use-case. We do so in the case of the European eGovernment Resource Network (eGRN) in which both authors are heavily involved.[5]

We first look at the application field of eGovernment and thereby set the context in which the eGRN project is situated. We then outline how eGRN uses specifically Topic Maps for the description of eGovernment Resources. We then show why centralized registries do not scale to the European eGovernment landscape.

In section 3 we present a prototypical technical protocol that does scale up to large, heterogeneous data sources. We do so alongside its implementation that is going to be elaborated and standardized in 2008 and 2009.

2.1   What is eGovernment?

eGovernment is "the use of information and communication technology [ICT] to support and improve public policies and government operations, engage citizens, and provide comprehensive and timely government services", to quote the widely accepted definition of the European eGovernment Society.

The term eGovernment itself is a child of the hype of the late 1990s, but IT usage in public administrations is, of course, much older. Since the 1960s a large number of eGovernment services were developed on all levels of government, European, national, regional and local levels — and many of these services are operational today. Services may directly interface individual citizens (e. g. regulatory reporting in countries where that is obligatory or applying for a new

---

[5] The exposition of eGRN in this section retakes some arguments from [10].

passport) or companies (e. g. VAT declarations, social security information) and other types of organizations. Most, however, are services from government bodies for other government bodies, be it as an internal service within an agency, be it cross-agency or, indeed, cross-administration.

Aside from eGovernment services, scores of other eGovernment resources, ranging from laws over specifications to forms, complement them.

2.2   The Premise Behind the eGRN

Legal and other regulatory requirements play a crucial role in governments. Almost all government services have a basis in some form of regulation.

The number of eGovernment services in Europe is not even approximately known; its number is certainly far in five, probably far in the six digits. Frequently, many different implementations realize functionally similar or identical services. Transparency on the situation promises both very significant savings and quality improvements through re-use and continuous improvement of existing solutions rather than unnecessary duplications. Only a standardized *description* of services and other eGovernment resources across Europe will permit to make eGovernment retrievable across Europe.

The Service Directive (European directive 2006/123/EC) and especially §6 on the points of single contact in each country[6] will push towards service interoperability within and, in the long term, across countries. Administrations must be able to find related services both within countries and across borders.

The main objective of the eGRN is thus to increase visibility for and better use of existing eGovernment resources for existing or foreseen projects and in a manner that can be managed and sustained without adding further costs to the maintenance of existing resources.

2.3   Linking eGovernment Registries

In the real world, there is rarely a single central registry for any given area of interest, and information on eGovernment resources are no exception. To use an example, each individual library usually maintains its own catalogue for the publications it owns. This makes sense: the individual library knows best about its collections and its specific needs, its new acquisitions and changes.

Restated on a more abstract level, the best knowledge about individual resources usually resides with the maintainers of those very resources themselves. It is something inherently decentral.

This is just as much the case for eGovernment resources: maintaining information about those resources is most effectively done by their actual providers — who very often also prefer to keep control over the provision and maintenance process of their data.

---

[6] The Service Directive specifies that "all procedures and formalities needed for access to [a service provider's] service" must be accessible through a point of single contact regardless of the number of organizations that are involved behind the scenes.

This desire for decentralized storage only seemingly conflicts with the need for central access to this information. Library catalogues are for many years now federated on different levels through common exchange formats and well-defined modes of cooperation — without compromising local data curation.

2.4   Description of eGovernment Resources

The eGRN is a child of the CEN/ISSS eGovernment Focus Group, an initiative within the European standardization body CEN that was created at the request of numerous public authorities and other entities with the goal to "map the various activities in the field of eGovernment standardization and to discuss a roadmap for the future" .

Following a Europe-wide call for experts, a three-person project team was selected in late 2005, including the two authors. In April 2006, in a first project phase supported by SUN Microsystems, the group created an ontology for the description of eGovernment services and captured a small sample set of service descriptions that is here shown in a revised version.

Standards and service descriptions that have been captured according this ontology to are displayed via a search mask build on top of NetworkedPlanet's commercial topic map engine TMCore.

The second project phase is partially supported by the European Union and EFTA through CEN. In it the project team focuses on the eGovernment standards report which will shed light on the role that standards and other types of specifications can play in eGovernment. The report is at present in final draft stage and open for consultation.

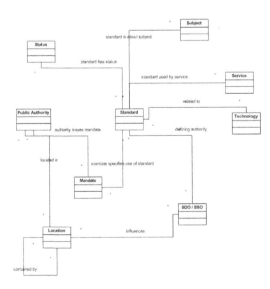

Fig. 2. Extract from the eGovernment ontology (Focus on standards, Source: Project team of the eGovernment Focus Group)

The second project phase has, however, seen another change, namely the introduction of a second topic map hub for data capture. The second engine, the Python-based open source implementation Py4TM, was developed by the University of Applied Sciences Worms and is primarily positioned for data capturing and storage.

3   Scaling Topic Maps

As postulated in section 1.4, one of the key prerequisites for scaling topic maps beyond the encyclopedic model is the decoupling of repositories and registries — something very close to the Topic Map paradigm —, but also the linking of individual registries to a large registry federation that is capable of exchanging information. The example of the eGRN shows a real-world setting where it is, indeed, paramount to conceptually transcend the closed world paradigm of the encyclopedia.

This per se is not something new — [11] and (to a much lesser degree) [12] both foresee multiple registry nodes. However, it has rarely been realized in practice, since most existing approaches mix separate aspects of registries that are quite orthogonal. In fact, the prerequisites for a successful linking of registries are case-specific agreements — standards — on three different levels of abstraction:

- on the technical level an exchange model for registry information
- on the semantic level an information model that define unambiguous labels for resources to clearly identify them across instances
- on the organizational level a collaboration model that defines who furnishes new or updated information

All three models are equally important for a functioning registry, yet can and should be specified separately. At present, little work has been done on the collaboration model, a notable exception being [13]. Such a model will have to be adapted to the business goals and organizational settings of each case. Similarly, the information model will reflect the basic structure(s) of domain knowledge, modelled through ontologies that are used for structuring information. In the case of the eGRN we have sketched them for standards and services in section 2.4.

3.1   The Exchange Model

In the remainder of this section we shall concentrate on the most technical part, the exchange model for registry information.

The existence of two registry nodes in eGRN implies right from the start the need to connect them to be able to automatically exchange data and to propagate changes across instances. However, while the topic map standard [14] specifies the rules for the merging of two or more topic maps including the merging rules for the "same" topic occurring in both maps (cf. [15], section 6), it explicitly does neither tackle the propagation of entries across topic map engines nor APIs for accessing topic maps or topic map fragments on actual implementations.

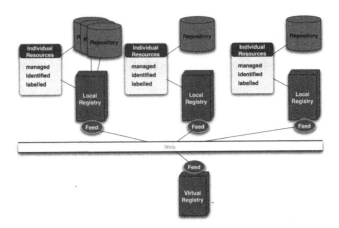

Fig. 3. Linking individual resources across registries (inspired by a diagram by Peter Brown)

Specifications for such linking formats are orthogonal to the merging rules of [14] (as in the spirit of separation of concerns they should be, cf. [10]).

That said, it is perfectly legitimate and in the spirit of the standard to complement it and the topic map data model [15] with precisely such mechanisms for specific application scenarios.

When designing the prototype currently in use between the two eGRN nodes, we thus specified a set of general and protocol design principles that our implementation should follow. For the general requirements, we agreed that

– data should be transmitted in a machine and human readable form
– the underlying transport format should be technology neutral
– the general architecture should fit naturally into a resource-oriented architecture (ROA) [16] of eGovernment resources
– the solution should in the mid- to long-term be able to support different collaboration scenarios, at least the publish-subscribe and peer-to-peer (P2P) scenarios

Starting out from these general premises, we settled on a few technical principles for the protocol design, namely:

– to use the ATOM [17] syndication format as a well-established and straightforward XML-based syndication mechanism
– to build transmitted entries out of XTM fragments
– to choose single transactions composed out of topic and associations fragments as information units
– to focus on instance data rather than updates to the underlying information model as the latter would in most cases entail updates to the nodes themselves

The Transaction Format: When designing an exchange format for the cross-implementation federation of our two eGRN nodes, we decided to build on ATOM for informing about additions and updates. The individual entries of the feed contain the XTM-rendering of the topic map constructs relevant to a specific transaction. Merging is done on the published subject identifiers that are obligatorily associated with all instance data.

A topic map engine then publishes an atom feed that contains an entry for each transaction (not each topic). Each item has two datetime stamps, one for its creation date and one (possibly identical) reflecting its last update, a unique key (GUID) and a link to the actual transaction object. The transaction object, in turn, consists of transaction elements encasing topic map fragments.

In the XML format all transactions are enclosed in `topicMapTransaction`-elements. They specify which of the three action types — the addition of new topic map constructs, and the update and deletion of existing ones — should be enacted and with regards to which topic map constructs:

- `addFragment`: The fragment is merged with the current map, new topics and associations are added
- `updateFragment`: all full topics listed are deleted along with all their assoc and replaced by the text in the fragment. Any topics not previously present are added
- `deleteFragment`: all topics are deleted with their associations, all associations are deleted

Collaboration Scenarios: Two different scenarios are conceivable for the setup of linked registries and hence the distribution of registry information. Registries can work according to the publish-subscribe model in which nodes publish ATOM feeds that are then consumed by other nodes, with feedback loops being explicitly forbidden (i. e. no node can act directly or indirectly as a subscriber to its own feed, the directed graph running up to each subscriber is a tree). They can, however, also be arranged as P2P nodes where that constraint does not hold true.

In the case of the eGRN we have for this moment settled on a straightforward publish-subscribe model with the Py4TM-based data-entry node acting as the publisher and the TMCore based search interface as the subscriber of the feed. This scenario is simpler than the full peer-to-peer protocol in that each subscriber keeps track of the news from a well-defined set of publishers.

In the P2P case the situation is more complex. Every node acts both as a publisher and subscriber of information, potentially for all the other nodes in the network. A given node X pulls all information on new transactions from his peers in the network and integrates it into its own registry, applying each change in sequence. Those transactions thereby show up in X's own ATOM feed of updates, additions and deletions that is published to all other peers. Those peers now carefully need to distinguish between transactions that are genuinely new to them and those that they have already consumed or that may even have originated from them.

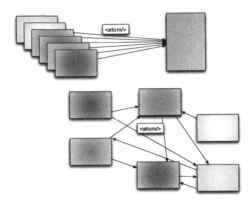

Fig. 4. Publish-subscribe and P2P exchange models for topic map transactions

In other words, every node needs to keep stock of all individual transactions that it has ever seen and applied. Each transaction must be uniquely identified using an ATOM entry with a universally valid and stable identification number.

These assumptions lead to a three rules for the processing of transactions across nodes:

- each atom item has a datetime stamp, a unique key (GUID) and a link to the transaction object
- each client engine is required to process the transaction and then add it to its transaction queue, including the original key to prevent duplicate processing of the same transaction and hence feedback loops. It must not generate new transaction for this update
- any updates that occur locally (outside the scope of this framework) must generate a new transaction and be assigned a new unique ID. Each peer keeps a log of transactions processed and created. It can then know not to reprocess a transaction that it created and then was re-published by another node

4   Steps into the Future

Currently, the services of the eGRN are prototypical and the protocol preliminary. Much especially in the case of P2P networks remains to be tested. An organizational model for collaboration must still be elaborated with major stakeholders in this scene.

To achieve these, the European Commission and EFTA have agreed to fund a follow up activity under the title *Universal Access to eGovernment Resources (UnivAc)*, again in the CEN framework. It is going to be formally kicked of on February 20th, 2008 in Brussels and will act as a sounding board for all three models. The results of this standardization activity are expected in Spring 2009, but the basic layout should be known well before that.

The topic map community will play a crucial part in shaping the exchange model which, we trust, can be a major step in the technical groundwork for a new, distributed and multifaceted encyclopedic model.

References

1. Arouet dit Voltaire, F.M.: De l'encyclopédie, Paris (1774)
2. Chartier, J.-L.A.: De Colbert à l'Encyclopédie. Presses du Languedoc (1989)
3. ATILF - Analyse et Traitement Informatique de la Langue Française: Encyclopédie, ou dictionnaire raisonné des sciences, des arts et des métiers (2007)
4. Le Alembert, J.R.d'., Malherbe, M.: Discours préliminaire de l'Encyclopédie. Librairie philosophique J. Vrin, Paris (2000)
5. Wikipedia: Porphyrian tree / arbor porphyriana (2007)
6. Küster, M.W.: Geordnetes Weltbild. Die Tradition des alphabetischen Sortierens von der Keilschrift bis zur EDV. Eine Kulturgeschichte. Niemeyer, Tübingen (2006)
7. Chambers, E.: Cyclopædia. Or, an Universal Dictionary of Arts and Sciences; Containing the Definitions of the Terms, and Accounts of the Things Signify'd thereby, in the Several Arts. James and John Knapton et. al., London (1728)
8. McLuhan, M.: Understanding media. Routledge, London (1964)
9. Mouchon, P.: Table analytique et raisonnée des matières contenues dans les vingt-trois volumes in-folio du Dictionnaire des sciences, des arts et des métiers. Panckoucke, Paris, Amsterdam (1780)
10. Küster, M.W., Moore, G., Ludwig, C.: Semantic registries. In: Tolksdorf, R., Freytag, J.-C. (eds.) XMLTage 2007 in Berlin, Berlin, pp. 21–36 (2007)
11. OASIS: ebxml registry services and protocols version 3.0. Oasis standard, OASIS (2005)
12. OASIS: UDDI — Universal Description Discovery & Integration. Oasis standard, OASIS (2004)
13. CEN/ISSS WS ADNOM: CWA 15526: The Establishment of a European Network for Administrative Nomenclature. CEN Workshop Agreement, CEN (2006)
14. ISO/IEC 13250: Information technology — SGML applications — Topic maps. ISO standard, ISO (2002)
15. ISO/IEC: ISO/IEC 13250-2:2006-08 information technology - topic maps - part 2: Data model. ISO standard, ISO/IEC (2006)
16. Richardson, L., Ruby, S.: RESTful web services. O'Reilly, Farnham (2007)
17. Nottingham, M., Sayre, R.: The atom syndication format. RFC 4287. Technical report, IETF (2005)

Convergence of Classical Search and Semantic Technologies – Evidences from a Practical Case in the Chemical Industry

Stefan Smolnik

Information Systems 2, European Business School (EBS)
Rheingaustr. 1, 65375 Oestrich-Winkel, Germany
stefan.smolnik@ebs.edu
http://ww.ebs.edu/is2

Abstract. With their expanding information assets and the growing importance of the knowledge factor, organizations are increasingly challenged to efficiently support knowledge management processes with appropriate structuring and retrieval technologies. Besides traditional information retrieval approaches, the use of semantic technologies like Topic Maps is also becoming more important. This paper compares the potential value of all these approaches, technological requirements, and applications. Thereafter, appropriate recommendations are derived for the most suitable choice of technology. In addition, early experiences from a practical case in the chemical industry provide further insights into the applicability of these recommendations in practice.

Keywords: Topic Maps, semantic technologies, information retrieval, text mining, technology evaluations, case study.

1 Introduction and Motivation

In today's office environments, work is affected by the escalating ratio of electronic information, which – mostly in the form of documents – is increasingly badly structured. The challenge is therefore no longer to obtain information, but rather to use it effectively and efficiently by specifically identifying relevant information from the large mass of information available.

Another substantial trend is that the resource knowledge and its management are gaining in respect of organizational meaning (see e.g., [VIN00, p. 3], [Rie04, p. 88f]). In the past decade, knowledge has become a crucial competitive factor as it has a large potential to generate value. Those organizations that are able to utilize their knowledge resource effectively and efficiently can thus obtain a sustainable competitive advantage from such value generation (see [Smo06, p. 3f]).

Information technologies – particularly technologies relating to information retrieval and structuring – address the challenges arising from information overflow and extensively support knowledge management to realize sustainable competitive advantages (see e.g., [Hec02, p. 2]). In the course of the World Wide Web's development into the Semantic Web, increasing attention has been paid to concepts and technologies based on semantic approaches. Besides the creation of the Semantic

L. Maicher and L.M. Garshol (Eds.): TMRA 2007, LNAI 4999, pp. 14–24, 2008.

Web, semantic technologies like Topic Maps can also be used to develop intra-organizational information systems that address the challenges described above.

Those responsible for intra-organizational information systems or knowledge management initiatives are confronted with a plethora of workshops and publications offering insight into semantic technologies' extensive potential for the creation of information and knowledge environments. The prerequisites required to develop this potential and the resultant costs often remain unmentioned. Consequently, this paper investigates information retrieval (statistical and linguistic searching, thus text mining), allocation of information attributes (structured searching, taxonomies, and thesauri), and information structuring (semantic technologies like Topic Maps) in respect of their specific potential value, requirements, and costs. In conclusion, the basic results are summarized in the form of recommendations for the most suitable choice of technology, which are further strengthened by the practical implications derived from first experiences gained from a case in the chemical industry.

2 Supporting Knowledge Transfer and Knowledge Construction

In the course of organizational value creation and delivery, the responsible persons frequently require information to complete assigned tasks, which can also be regarded

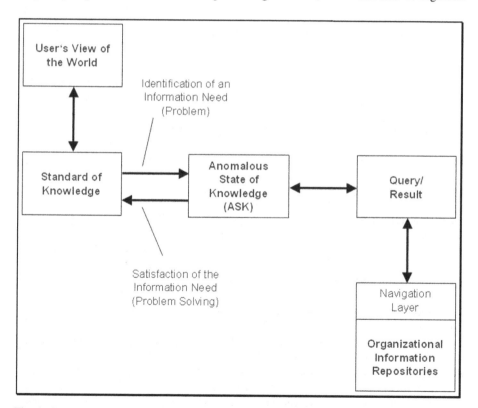

Fig. 1. Cognitive situation of the user of organizational information repositories (according to [Teu96, p. 9])

as a lack of knowledge about a specific problem. These persons thus reveal inadequate or no problem-solving skills at all. Generally speaking, Belkin describes this kind of situation as an "anomalous state of knowledge" (ASK), which describes an individual's problem situation from a cognitive viewpoint, revealing a lack of knowledge with regard to a section of reality (see [Bel80, p. 137], [Teu96, p. 7 f.], as well as figure 1).

Consequently, the responsible persons' mental model is incomplete with regard to the given problem and its solution, which motivates them to take their information need into consideration and satisfy it. The information need should not be a priori specifiable, i.e. the responsible persons must not exactly know what they are looking for. Within the context of knowledge transfer, those responsible for a task take on the role of the recipient. They decode, study, and adapt relevant information objects found to satisfy the need for information. Consequently, these information objects are integrated into the existing mental model, i.e. new knowledge is constructed.

The support of organization employees by means of powerful tools that search, navigate, and explore organizational information repositories is a prerequisite for knowledge retrieval and application. These tools simplify and support both knowledge transfer and construction, as well as helping to identify relevant information objects and, eventually, constructing new knowledge.

3   Technology Evaluations: Potential Value, Requirements, and Costs

It has been suggested that the way of using metadata should be considered the classifying characteristics in respect of the evaluated technologies below. However, the technology presented first – information retrieval or full text search – does not employ metadata. Conversely, the technologies introduced subsequently – structured search and semantic technologies – use flat or hierarchically organized metadata, or metadata organized in a network.

3.1 Information Retrieval

Traditional information retrieval technologies employ full text searching (also: free searching, free text searching, and searching by means of keywords) in which arbitrary search terms are applied to convey information requirements (see [Kre04, p. 21ff]). Full text search systems produce details about information objects whose index terms are similar to the search terms. Google is a typical example of such a system (http://www.google.de). Full text search systems are characterized by a very high search performance and scalability, and a low implementation cost. In contrast to structured search and semantic technology approaches, no explicit knowledge models are required for information retrieval technologies. Consequently, the modeling costs remain low. A further advantage of information retrieval or full text search technologies is the linking of index information with other details such as relevance.

A requirement for using full text search systems is the availability or the implementation of specific interfaces for the indexing of heterogeneous information assets. Intellectual efforts (e.g., finding matching search terms for existing content) are replaced by search activities. The costs of the search for information are thus mainly borne by the user. A targeted search of large information assets requires the user to have a high level of technical and linguistic competence. Furthermore, full text search systems based on statistical methods only retrieve documents that specifically contain the words/partial words included in the search query; there is also no terminological check of the search terms' syntax (see [Fer03]). These statistical approach limitations can be avoided through the use of thesauri and control systems, which do, however, lead to additional costs. Finally, the disk space required by indices, which can comprise up to 80% of the indexed file's disk space, should not be ignored.

3.2 Assigning Attributes to Information

In a structured search (also called an attribute-based search), the author and the user employ the same syntax for indexing and searching. This syntax makes use of a common classification with descriptions provided by selection lists ("controlled vocabulary"), i.e. the actual search is done by means of metadata (also called attributes and descriptors). Consequently, by selecting classification characteristics, the author's context is clear to the user and/or can be reconstructed from the employed terms. Furthermore, a higher quality knowledge model, based on an explicit and intellectual knowledge model, is also obtained, which the full text search systems presented above cannot present. In addition, a structured search produces a smaller number of results, allowing the speedier identification of relevant information objects.

Besides the described advantages of a structured search two challenges need to be mentioned. On the one hand, search queries are limited to content based on specified topics, thus preventing the free exploration of information assets. On the other hand, the employed controlled vocabulary must be maintained (terminology management), i.e. information objects must be provided together with the appropriate metadata that are stored with them. However, the organizational use of these terms can change over time, requiring reclassification. Alternatively, outdated terms can be mapped onto new terms (e.g., with a special thesaurus).

3.3 Information Structuring

Semantic technologies like Topic Maps structure information assets in the form of networks. Similar to an ontology, Topic Maps help to structure and describe an application field (knowledge domain) that all persons and applications jointly utilize (see e.g., [ZSS01]). Relevant terms and their relationships are defined precisely for a specific application field. Therefore, information objects are not juxtaposed separately (as in the case of information retrieval or full text search) or hierarchically (as in the case of a structured search), but are linked. Information is thus characterized in more detail by means of statements about itself and about its relationships.

Semantic networks presented by Topic Maps enable the identification and explication of various contexts that were not previously apparent. By linking related themes, Topic Maps support human thought processes (humans think by means of associations) (see [Smo06, p. 61]). These processes are specifically supported by this

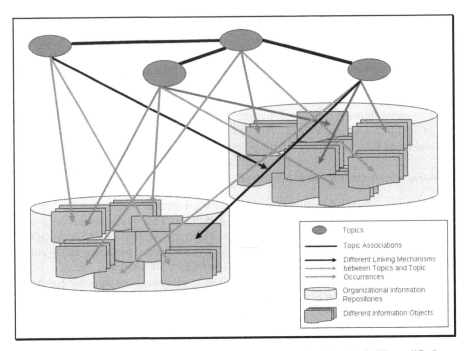

Fig. 2. Information structuring through the use of topic maps ([Smo06, S. 58] modified)

technology's distinctive navigational and explorational capabilities. A further advantage of Topic Maps is the integration capability of heterogeneous information assets.

An important requirement for the employment of this technology is the production or generation of a semantic network, i.e. of a specific topic map. It is therefore possible to generate a topic network, as well to link these topics with information objects (see figure 2). Expenditure on the design, generation, and maintenance of these structures is mostly considerable, since it normally requires manual procedures. Experts are, for example, required for the applied technology, as well as for modeling the addressed knowledge domain. Depending on the information assets' degree of structuring, information objects also have to be annotated. The use of semantic technologies frequently means a new search and navigation paradigm for users who are normally only accustomed to flat and hierarchical information structuring.

4 Recommendations

The results of the technology evaluations are summarized in figure 3. Depending on the degree of structuring and the size of the information objects' content, the degree of maturity, the requirements and implementation costs, the quality of the search results, the search and retrieval performance, as well as the heterogeneous information assets' integration capability are estimated in respect of the technologies examined.

Extend of Content			Full Text	Text + Metadata
Degree of Structure			weak	semi
Organization of Metadata	Technology			
	Full Text Search			
		Statistical Approaches	M: ●	
No Metadata			R: ○	R: ○
			Q: ◉ - ●	Q: ○ - ●
			P: ●	P: ●
			I: ○	I: ○
		Linguistic Approaches	M: ◉	
			R: ◉	
			Q: ◉	
			P: ◉ - ●	
			I: ○	
	Structured Search		M: ●	
			R: ●	R: ◉
Flat or Hierarchical Organizations of Metadata			Q: ◉	Q: ●
			P: ●	P: ●
			I: ○	I: ○
	Taxonomies & Thesauri		M: ●	
			R: ●	R: ◉
			Q: ◉ - ●	Q: ◉ - ●
			P: ●	P: ●
			I: ○	I: ○
	Topic Maps		M: ○	
			R: ●	
			Q: ●	
			P: ●	
Network-like Organization of Metadata			I: ●	
	Ontologies		M: ◉	
			R: ●	
			Q: ●	
			P: ●	
			I: ●	

M - Maturity	● high / ◉ average / ○ low
R - Requirements for and Efforts of Implementation	● high / ◉ average / ○ low
Q - Quality of Search Results	● high / ◉ average / ○ low
P - Search/Retrieval Performance	● high / ◉ average / ○ low
I - Heterogeneous Information Repositories' Integration Capability	● high / ◉ average / ○ low

Fig. 3. Technology evaluations

A series of fundamental recommendations and conclusions can be derived from the results:

- In respect of widely-used information objects (e.g., MS Office documents, PDF and HTML files), full text search systems produce good to very good search results. This is especially true in the case of the combined use of statistical and linguistic approaches. Owing to the proven algorithms and rule systems, as well as available dictionaries for general vocabulary, requirements and costs are low. Full text search system should specifically be used if not only a simple and quick to implement search and retrieval solution is preferred, but also if the search will be mainly focused on concrete facts, information, and information objects (the user can specify his information need relatively clearly).

- The quality of the search result, as well as the search and retrieval performance is higher in structured searching and in thesauri/taxonomies than when full text search systems are used, as the desired information objects can be better targeted. The use of these technologies is specifically recommended in cases where the information assets and objects underwent a suitable pre-structuring. However, the costs of the structuring (controlled vocabularies/thesauri/taxonomies, metadata) and the expected gain in quality should be compared critically.
- The requirements and costs are the highest when semantic technologies like Topic Maps are used. Suitable topic networks are usually not available and it is often necessary to suitably prepare existing information objects (e.g., through metadata and annotations). Semantic technologies do, however, offer an advantage if the knowledge domain that is to be modeled is manageable and relatively sharply definable. Their use is particularly recommended for explorative searches and fuzzy retrieval queries, i.e. when users are not sure what they are searching for, or when new relationships and as yet unknown information have be found.
- Topic Maps have a distinct integration ability with regard to heterogeneous information assets. Conversely, other technologies can be extended with suitable connectors and/or federated search approaches to handle distributed heterogeneous information assets. Normally, this means an increase in costs (e.g., the implementation of interfaces, consolidation of result sets, various terminology sets), as well as a decrease in the quality of the search results and/or search and retrieval performance (e.g., larger result sets, lacking relevance considerations).

When choosing technologies for structuring and information retrieval, the planned usage must be examined in detail:

- Which processes are to be supported?
- Are pre-structured information assets and/or information structures available?
- What types of information requirements will have to be satisfied?

Answers to the above questions will indicate the technology that can provide the best relationship between the potential value, requirements, and expenditure.

5 Evidences from a Practical Case in the Chemical Industry

The presented challenges and trends also apply to the information and research department of a major German chemical firm, where extensive research is done to support scientists' activities and information systems are offered to internal customers to configure their research activities more efficiently and to ensure that these activities are of a high quality.

The following described evidences have been derived from early experiences gained from a practical case in this department, during which the benefits and challenges were explored of deploying and applying both text mining and semantic technologies to the very specific and heterogeneous information landscape (consisting, among others, of patent and scientific literature databases). In the course of these studies, a technology matrix was constructed (figure 3 is a generalized version of a small part of this matrix). Use cases for the knowledge domain Skin

Cosmetics as well as software requirements for adequate information systems based on semantic technologies were also formulated. Furthermore, vendors of information systems based on text mining or semantic technologies were interviewed and invited to demonstrate their information systems.

The concepts, methods, and processes of text mining and semantic technologies as discussed above can be compared and evaluated by means of several criteria:

- Basic principle of information (pre-)processing: Text mining and semantic technologies can be distinguished by means of their underlying approaches to pre-process and process information. Whereas automated document processing receives priority in text mining, an intellectual or semi-automated knowledge modeling is carried out when using semantic technologies. Depending on the addressed knowledge domain, these two approaches lead to different results. For example, when automated text mining methods are used in colloquial English texts, they provide very good results. If the text language changes and specificity grows, the effort in terms of manual modifications increases, also within text mining methods. The attractiveness of semantic technologies therefore increases, especially if there are already pre-structures such as a thesaurus or taxonomy for the knowledge domain. However, semantic technologies might be comparatively easily deployed in the given example, as statistic and linguistic methods achieve good results when creating knowledge models.

- Knowledge models and knowledge modeling: Text mining, as well as semantic technologies rely on explicit (knowledge) models and knowledge modeling. In this context, text mining systems normally use standard models such as thesauri, encyclopedias, and rule systems that are able to indicate, for example, the grammar of a language or enable the identification of entities (like a city or persons' names). There are such standard models for common vocabulary, for special branches or industries (e.g., the pharmacy and medicine industries), and for special knowledge domains (e.g., for the extraction of finance, business, and stock market data, as well as of the data of mergers & acquisitions, joint venture, and innovation). The increasing specificity of the language to be processed and of the knowledge domain leads to a greater need for manual extension or even a new conceptual design for the used models. Another characteristic of these models is that – once they have been designed and implemented – they are often stable and rigid, i.e. there is usually no dynamic modification to meet new requirements that might occur in business. In contrast, semantic technologies implement a flexible, intellectual knowledge modeling to meet the specific requirements of the knowledge domain, often regardless of specific document collections.

- Type and semantics of relations: Text mining systems make use of relations – specified by standard models – between concepts that these models also provide. Typical examples are relations between [person] and [company] or between [person]/[company] and [market]. Furthermore, relations defined in encyclopedias or thesauri are often "simple" (i.e., semantically poor), such as those of a hierarchical nature. Users of such systems are therefore limited to the given concepts and their relations. In contrast, semantic technologies rely on relations in the semantic network, i.e. in the explicit knowledge model. Those relations can basically be freely modeled and are therefore semantically rich (example: "[process A] creates [component B]"). Thus, according to the defined requirements, an ideal knowledge model can be designed for a specific application scenario, whereas text mining systems revert to standard models (which are vendor specific in many cases).

- Efforts: In the context of accessing information assets, different kinds of efforts can be differentiated. In the case of knowledge-oriented text mining, the various systems have to offer the required intelligence and provide the necessary information accessing capabilities. In the case of method-oriented text mining, users have to cope with exploration efforts for the identification of relevant contexts. When semantic technologies are used, developers of respective information systems and users have to cope with specific efforts such as for the designing of explicit knowledge models, i.e. semantic networks. These efforts should not be underestimated, as various practical examples, as well as experiences during this case demonstrate. The usefulness of an information system based on semantic technologies, as well as the users' satisfaction correlates directly with the quality of the knowledge models.
- Requirements: A basic requirement that applies to both text mining and semantic technologies is the effort that document pre-processing requires. In the case of text mining, a pre-processing of the relevant documents occurs, which is mainly based on linguistic methods. However, semantic technologies need an adequate annotation of information objects, as well as a link between the concepts and the information represented.
- Heterogeneous information assets: A further aspect is the processing of heterogeneous information assets. In this case, semantic technologies achieve integration on a semantic level. In contrast, text mining is used to extract certain information. Some text mining applications process only a single document; other applications address document collections, but categorize or arrange them only and therefore do not achieve "real" semantic integration.

5.1 Basic Findings

The evaluation of text mining and semantic technologies allows some key statements and conclusions:

- When a comparison is made of text mining and semantic technologies, it is possible to distinguish two basic methods: With the help of automation, text mining tries to automate all information processing efforts for the user. In contrast, semantic technologies focus on the advantages (especially the quality of the search) of the intellectual modeling of knowledge networks.
- However, there is an increasing convergence of information systems based on text mining and those based on semantic technologies. Standard models used by text mining systems often use simple ontologies, whereas semantic methods focus on text mining methods to not only support the efficient modeling of knowledge, but also to link semantic network's concepts with concrete information objects. There is a specific convergence between text mining and semantic technologies regarding document pre-processing. The similarities and differences between information access and processing by means of text mining and semantic technologies are summarized in figure 4.

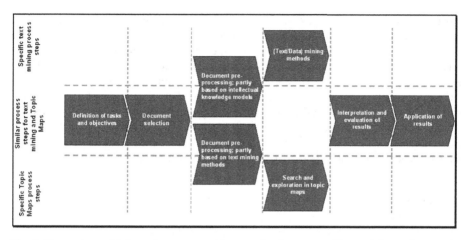

Fig. 4. Comparison of information access and processing by means of text mining and semantic technologies

- As a specific vocabulary is often used and specific knowledge domains are often addressed in an organizational context, the need to model (knowledge) structures intellectually, even in the context of text mining, becomes greater. In the context of a specific application scenario, the question arises regarding how much additional effort is required to fulfill the given requirements directly with the help of semantic technologies.

6 Conclusions and Areas for Future Research

In this paper, we have evaluated methods for information access and/or retrieval, structured search, and semantic technologies. Besides introducing basic paradigms, concepts, and methods, we have also indicated their requirements, efforts, benefits, and challenges. In addition, some first results from a practical case in a chemical firm have been presented. In the context of this case, text mining and semantic technologies have been specifically investigated. Even though the technologies differ – text mining employs (fully) automatic information access while semantic technologies focus on intellectually modeling knowledge structures –, an increasing convergence of these technologies is noticeable. Text mining applications demonstrate a growing reliance on ontologies or similar approaches, while applications based on semantic technologies support the knowledge modeling process through text mining tools. Consequently, the respective processes for information access and processing are increasingly converging.

One major area for future research is gaining more detailed insights and practical experiences by applying text mining and semantic technologies in a typical environment. Besides the conceptual considerations covered by this study, the described convergence of the two technologies makes it necessary to determine practical impacts when information systems are employed that are based on the these technologies, i.e. during a proof of concept project in a test environment. In addition, business cases should be defined for implementing, deploying, and running such information systems to gain further economic insights.

According to the Gartner Hype Cycle for emerging technologies, the organizational usage of semantic technologies is about to reach its "peak of inflated expectations" (see [PG06]), i.e. corporate semantic technologies will next reach the "trough of disillusionment." In addition, Gartner predicts that semantic technologies will not be commonly available in organizations within the next five to ten years. In spite of this rather conservative judgment, it is advisable to actively observe semantic technologies' further development and gain practical experiences to identify and realize chances and benefit potentials that may arise from them.

References

[Bel80] Belkin, N.J.: Anomalous States of Knowledge as a Basis for Information Retrieval. The Canadian Journal of Information Science 5, 133–143 (1980)

[Fer03] Ferber, R.: Information Retrieval – Suchmodelle und Data-Mining-Verfahren für Textsammlungen und das Web. Dpunkt-Verlag, Heidelberg (2003)

[Hec02] Heckert, U.: Informations- und Kommunikationstechnologie beim Wissensmanagement – Gestaltungsmodell für die industrielle Produktentwicklung. Deutscher Universitäts-Verlag, Wiesbaden (2002)

[Kre04] Kremer, S.: Information Retrieval in Portalen – Gestaltungselemente, Praxisbeispiele und Methodenvorschlag. Institut für Wirtschaftsinformatik, Universität St.Gallen, St.Gallen (2004)

[PG06] Pettey, C., Goasduff, L.: Gartner's 2006 Emerging Technologies Hype Cycle; (November 25, 2007) accessed online: Gartner, Inc., (2006), http://www.gartner.com/it/page.jsp?id=495475

[Rie04] Riempp, G.: Integrierte Wissensmanagement-Systeme – Architektur und praktische Anwendung. Springer, Berlin (2004)

[Smo06] Smolnik, S.: Wissensmanagement mit Topic Maps in kollaborativen Umgebungen – Identifikation, Explikation und Visualisierung von semantischen Netzwerken in organisationalen Gedächtnissen. Shaker, Aachen (2006)

[Teu96] Teuber, T.: Information Retrieval und Dokumentenmanagement in Büroinformations systemen; Unitext-Verlag, Göttingen (1996)

[VIN00] Von Krogh, G., Ichijo, K., Nonaka, I.: Enabling Knowledge Creation – How to Unlock the Mystery of Tacit Knowledge and Release the Power of Innovation. Oxford University Press, New York, USA (2000)

[ZSS01] Zelewski, S., Schütte, R., Siedentopf, J.: Ontologien zur Repräsentation von Domänen. In: Schreyögg, G. (Hrsg.) Wissen in Unternehmen – Konzepte, Maßnahmen, Methoden, pp. 183–221. Erich Schmidt Verlag, Berlin (2001)

A Citizen's Portal for the City of Bergen

Lars Marius Garshol

Bouvet ASA, Oslo, Norway
larsga@bouvet.no
http://www.ontopia.net

Abstract. In early 2007 the city of Bergen, the second largest city in Norway, launched a new citizen's portal based on Topic Maps. This paper describes the project, the technology used, and some lessons learned from the project.

1 Introduction

In February 2007 the city of Bergen launched a new citizen's portal, replacing an existing, older portal. The new portal is the result of a strategic decision by the city administration to make the portal the future main channel for communicating and interacting with the citizens. This requires reengineering a number of business processes inside the administration to move them online, and the expectation is that this will yield substantial efficiency benefits.

Bergen is Norway's second city, located on the western coast, and famous for its constant rain, Hanseatic history, and well-preserved architecture. The city has 244,000 inhabitants, and 18,000 employees working in the city administration. The 2008 budget of the city administration runs to 1.6 billion EUR.

The project started in late 2004 with a year or so of preparatory work, such as developing the vision of what the portal should be, and studying the requirements and expectations the citizens had for the portal. Based on this, the city administration came up with a number of goals for the portal:

- It must be easy for citizens to find relevant information in the portal, whether by navigation or search.
- To support the previous goal, information should be organized by subject, and not, as in the previous portal, by the internal structure of the city administration.
- The portal must provide information workers with a flexible set of tools for building pages and portal structure.
- Data integration must be flexible and as automated as at all possible.

What's called *phase 1* of the project started in late 2005 with work on selecting technologies and vendors. In early 2006 Topic Maps, with Ontopia as the technology vendor, and Bouvet as the systems integrator, were chosen. This was done based on evaluation of a Topic Maps-based solution concept, consisting of drafts of the ontology, the data flow, an administrative interface, and an overall

L. Maicher and L.M. Garshol (Eds.): TMRA 2007, LNAI 4999, pp. 25–35, 2008.

concept. The concept was developed by Ontopia and Bouvet in collaboration with the customer.

Once the decision was made, the interaction design, which had already been developed in late 2005, had to be updated to take into account the use of Topic Maps. The portal was then implemented (both user interface and data integration), tested, and, on 6 February 2007, deployed at http://www.bergen.kommune. no. By then the city administration had spent approximately 5 million Euro on the project, of which only a fraction (less than 25%) had been spent on the Topic Maps part of the project.

In phase 1 only the basic platform and framework for the portal with a limited number of initial services had been built. Immediately after the portal went live *phase 2* started, aiming to add many new services, and also to make a large number of new departments in the city administration start publishing their content in the portal. These activities have a large impact on the organization, and make up a large proportion of the overall project cost. In addition, personal pages for each citizen will be added where citizens can see data registered about them and the progress of processes they are involved in.

Phase 2 is scheduled to end in December 2007, and to be followed by *phase 3*, the contents of which are currently not known.

2   Ontology

The ontology was developed during a series of workshops with participants from the portal project as well as representatives for the main systems used as data sources. The resulting ontology was created based primarily on the data already existing in various sources, but also on the requirements of the city administration.

The ontology can be divided into the following main parts:

- **Employees and departments.** This part has all city employees with contact information, all city departments with contact information, employment relationships, and parent/child relationships for all departments.
- **Taxonomy.** This is a taxonomy of subjects used to classify other information in the portal, such as articles, departments, and services. It originates outside the city administration (see 3.1 on the facing page).
- **Services.** This part has descriptions of the services provided by the city administration, as well as relevant forms, whether online or paper-based. There are also connections to relevant departments.
- **Articles.** This part has the articles published in the content management system (CMS), as well as some of the organizing structure from the CMS. Articles have "is-about" associations with nearly all other topics in the topic map to express what the articles are about.

An outline of the ontology is shown in figure 1 on the next page.

Currently, the ontology is being extended to include a calendar of events, the political organization of the city administration, and important locations in the city (like concert halls, sports venues, etc).

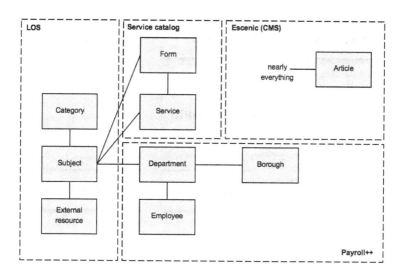

Fig. 1. The ontology

3   Technologies

The portal runs on Oracle Portal, but this is just the portal container. In addition, it uses the Ontopia Knowledge Suite (OKS) for the Topic Maps support, and the Escenic Content Engine as the CMS for articles and other content. An integration between Escenic and the OKS was built for the project by Escenic and Ontopia, and is described in 3.2 on the following page.

3.1   Data Sources

A number of external systems deliver data to the portal by various means. The systems involved are:

- **Fellesdata.** This literally means "Common Data" and is an internal project in the city administration to create a common data warehouse integrating data from many different sources, to be used by any client system within the city administration. Data is loaded from here into the topic map using DB2TM [Garshol06a].
- **Dexter and Extens.** A school and kindergarten management system. Information about kindergartens is loaded from here into Fellesdata.
- **Agresso.** A payroll system. Information about departments and employees is loaded from here into Fellesdata.
- **Local Service Catalog.** A system for writing descriptions of the services provided by municipal governments in Norway. Information about services is loaded from here into Fellesdata.
- **LOS.** A common taxonomy for classification of information from Norwegian municipal governments published by Norge.no as a topic map [Norge.no]. Data is loaded from here into the topic map using TMSync [Garshol06b].

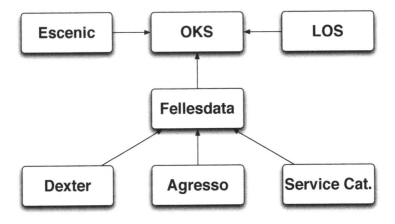

Fig. 2. Dataflow

It would have been technically possible to integrate Dexter, Agresso, and the Service Catalog directly with the topic map, but to use Fellesdata to integrate data from multiple sources was more in line with city administration policies. It was also easier for the project to get all its data from a single source, and as deadlines were tight this was preferred.

3.2   CMS Integration

The CMS used is Escenic Content Engine, developed by Escenic, a Norwegian company. The Escenic CMS is widely used in the media industry, but also used by other customers, like the city of Bergen. Escenic has a large international customer portfolio, boasting users like The Times (the London newspaper), The Independent, and the Dutch newspaper de Telegraaf. The integration with the OKS was developed for the city of Bergen project by Escenic and Ontopia in collaboration and is now available as an Escenic module.

The internal structure of Escenic revolves around sections and content objects, which are primarily articles, but can also be multimedia of various kinds. Sections are parts of a web site into which content objects can be *desked*, that is, added. The basic principle of the integration is a data integration which creates topics in the OKS for Escenic articles and sections. A standard Escenic ontology is defined as part of the module, and contains topic types like `ece:article` and `ece:section` and association types like `ece:article-in-section`, and so on. The integration is configurable, so that users can control which articles and sections are included in the topic map, and which are not.

Article types (like FAQ, news article, Q & A, ...) are mapped into topic types in the topic map, as subtypes of `ece:article`. Each article becomes a topic with its article type as the topic type and its title as the topic name. Some other fields, such as who created the article, when the article was created, and its workflow state are also included, but the main text of the article is not included. The section structure and linkages between articles and sections are also included.

The integration consists of three parts:

– A component which receives events from the Escenic API and updates the topic map accordingly, creating and deleting topics for articles and sections etc.
– An integration of Ontopia's Ontopoly Topic Maps editor into the WebStudio user interface to Escenic, which allows users to associate articles with topics inside the Escenic interface, making publishing and annotation a single operation. The annotation interface is the Ontopoly instance topic editing screen.
– A tolog module containing a predicate supporting full-text search of Escenic articles. This is used to send full-text queries to Escenic and produce topics representing found articles, and can be combined with other predicates to implement a more complete search of the topic map.

The existing JSP libraries of Escenic and OKS worked together out of the box, and so no changes were necessary to allow a single web application to produced from both Escenic and OKS.

3.3   Technical Architecture

The technical architecture is shown in figure 3. Web traffic is first received by a load balancer (not shown), which divides traffic between the two proxy caches, both of which run Oracle Webcache. These cache most of the pages, although not the search pages, in order to reduce the load on the servers behind them.

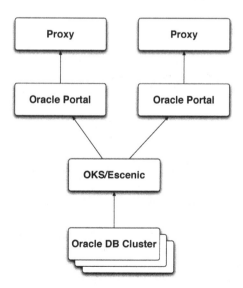

Fig. 3. Technical architecture

The Oracle Portal servers run the portal itself and render the structure of the individual pages, but do not actually execute the portlets that populate this structure with content. These are instead rendered by the Oracle Containers for Java (OC4J) server that runs both OKS and Escenic.

Finally, there is a cluster of Oracle database servers which are used by both Escenic and OKS to store content and the topic map.

The two-pillar structure of the architecture has been set up primarily to ensure that the solution has more than a single point of failure, but also to ensure that it scales at times of peak traffic. There is only a single active OKS/Escenic server at any given time, but another is kept on standby in an active/passive failover configuration. For performance figures, see 5.3 on page 34.

4   The Portal

The old portal for the city administration was organized by department, which meant that if a bar owner wanted to apply for a license to serve liquor in a bar, they would have to navigate to the Department of Finance, since that is the responsible department. This was seen as suboptimal, and so the decision was made to give the new portal a more logical form of navigation. This was part of the rationale for choosing Topic Maps.

In the current version, the two main points of entry into the portal are navigation through the LOS taxonomy and search. In general the portal follows the topic-page pattern already known from many Topic Maps-based portals, where most pages represent a single topic, and show all information about that topic. Associations are rendered as links to other topic pages.

4.1   Interaction Design

The front page of the portal as it was on 2007-12-03 can be seen in figure 4 on the next page. The leftmost column consists of an A-Z list of links into categories and services, followed by a menu built by the editors. In the middle column is a search field, followed by a view of the two topmost layers of the LOS taxonomy, and below it a list of news articles desked to the front page.

In figure 5 on the facing page is shown the page for the LOS topic "Children and families". On the left is the search box and the A-Z navigator. In the middle is a breadcrumb trail, the heading, and then links to the child categories, followed by a list of relevant articles. On the right-hand side are first related services and forms, followed by related resources on the web from Norge.no (publishers of LOS).

The search pages (not shown for reasons of space) use a faceted search approach, allowing search results to be filtered by topic type, LOS category, and geographic affiliation. The result is a very powerful search tool, although it has been criticized for having a too complex layout.

Fig. 4. Portal front page

Fig. 5. LOS page for "Children and families"

4.2 The Portal Framework

The portal is built from portlets, like Escenic's ContentPortlet (which shows either an article or a section) and the OKS's related topics portlet (used to show the associations of a single topic) and the OKS's menu portlet. In addition, numerous custom portlets have been built reflecting the more specialized requirements of the interaction design.

The custom portlets are developed using JSP, mainly based on the JSP taglib provided by the OKS. Some additional tags have been developed specially for the project. The most important of these is a tag which produces a link to any topic based on a URL pattern stored in the ontology.

4.3 Maintenance

The reponsibility for the portal is held by a small department under the section for city development. The department has a manager, and three groups under that manager:

- The organization group, primarily staffed by external consultants, which is responsible for the business process reengineering required to move services into the portal.
- The technology group, which is responsible for the technical implementation of the portal and the data integration. It is partly staffed by external consultants and partly by internal employees.
- The information group, consisting of the portal editor and an interaction designer.

In addition, some 50 editors and authors in various departments publish content into the portal, and some have been given responsibility for the content in various parts of the portal. This number is expected to increase sharply as more departments start publishing in the portal towards the end of phase 2.

5 Experiences

This section describes the main experiences from the project.

5.1 Evaluation

Overall, the city administration considers the project successful and is proud to have built such a modern and forward-looking portal. This is not to say that all aspects have been equally successful, however. The least successful part of the project, in the eyes of this author, is probably the interaction design, which was inherited from before the decision to use Topic Maps was made, and which has never been property reviewed. Instead, changes have been made piecemeal, and the result is an interaction design that is partly good and partly very poor. In short, insufficient attention has been given to this aspect of the portal.

Project managers, customer representatives, and developers have described the project as giving the feeling of being "the first people on the moon." They have ascribed this to their unfamiliarity with Topic Maps, and have felt that not only is the learning curve quite steep, but there are also few sources of information about Topic Maps.

The tools used were less mature than, say, tools for developing with relational databases, and a number of components were either developed specially for the project, or first being deployed in this project. This meant that not everything the project needed was supported out of the box, and some tasks were harder to accomplish than they needed to be. For example, patching and modifying the topic maps and moving patches between test and production environments is somewhat awkward. Stability and bugs have not been serious issues, however.

A major issue with the portal from the point of view of users has been that the portal lacks much of the content that users seek. This is partly because not all departments publish via the portal yet, and partly because the project's knowledge of what users actually want was imperfect. Steps are being taken to solve both of these issues at the moment.

The project is large, with many participants from various branches of the city administration, as well as a number of external companies. This complicated the project considerably, and slowed down decision-making, but it's not clear whether this could really have been avoided.

Overall, the project did meet the customer's goals, in that the portal is highly navigable and searchable, and the portal framework really has provided the flexible toolkit the editors asked for. The data integration also works as previously advertised, and the only manual steps are caused by omissions and mismatched structures in the source data.

The developers were also happy with the choice of Topic Maps, and felt it was a more intuitive way of working with the data than traditional technologies. They've also stated that it's "fun to work with Topic Maps".

5.2 Lessons Learned

A number of useful lessons have been learned from the project, one of which is that for the portal to really satisfy the needs of users three different parts of the project all need to succeed:

- The ontology design. That is, the ontology really must match what users want and how they think.
- The interaction design. Even if the ontology proves a perfect match it's entirely possible for the interaction design to confuse users.
- The information design. That is, how the information to be conveyed to users is divided into articles and multimedia attachments. Here, again, it's possible to to make choices that confuse users, for example by running against the grain of the ontology design, or simply by being suboptimal.

Another lesson is that typically editors will need finer distinctions in the ontology than users do. In the portal categories, that is, nodes in the LOS taxonomy, can be of four different kinds:

- "Everything about"-page. These pages gather content on subjects that are of permanent importance, and about which there is a sizable amount of content.
- "Interest area". These are very similar to "everything about" pages, but are more temporary in nature, and there is less content about them.
- "Subject". A leaf node in the LOS taxonomy that has not been elevated to the status of either "everything about" or "interest area". Content gets attached to these as it is added, but there is no concerted effort to coordinate the content under these nodes.
- "Theme". The same as "subject", but "themes" are non-leaf nodes. A business rule in the LOS taxonomy states that these are not to have content attached directly, in order to simplify changes to the taxonomy.

These distinctions are all important, as they reflect different editorial policies and usages of the different kinds of categories. However, they are only important to editors, and users simply find this terminology baffling. Therefore, it is important for the portal to allow editors to make these distinctions, but to also hide them from users, so that to users these are all simply "categories".

The decision to organize content by subject rather than according to the organizational structure has caused some complications. It is no longer obvious which department is responsible for what sections of the portal, and making the decision about responsibility can be difficult in cases where the responsibilities of departments overlap.

Another lesson is that seeing the portal from the perspective of the users is very hard, and knowing what the users seek is in many cases just not possible. In this case, the users are roughly 250,000 people, compounding the problem. The project has found search log statistics to be an invaluable guide to user requirements, and that running popular queries on the portal to study the results helps judge the effectiveness of the portal. Some of the ontology changes and many of the information design changes in phase 2 were motivated by insights gained from studying search log statistics.

Finally, it should not be taken for granted that editors and authors of content have the skills necessary to categorize their content against the topic map. One issue is understanding the ontology and how to use it, but this is often not the biggest challenge. Most non-specialists simply do not understand categorization, and how to choose the best topics, the right number of topics, and how to avoid choosing redundant topics. Training is almost always necessary in order to improve the quality of categorization.

5.3   Performance

From the portal went live in February 2007 to May 2007 21.7 million requests were received, which corresponds to 189,000 per day, or 2.2 per second. Traffic has obviously been higher at peak times. 55% of requests took no measurable time to answer, 83% took less that 0.1 second, and 95% less than 1 second. So performance has obviously been satisfactory.

The topic map is quite small, with 7200 topics, 16,000 associations, and 17,000 occurrences. Escenic has 1140 articles. Given these numbers, it would have been surprising if the portal had struggled with performance problems. As more departments start publishing into the portal during phase 2 these numbers are likely to increase significantly, but it is too early to know what impact, if any, that will have on performance.

6   Conclusion

The portal has currently been in operation for about 5 months, and the city administration is overall pleased with what has been realized so far. The main complaint is that the interaction design is not good enough, and work is in progress to correct this.

The main task for the portal project at the moment is to add more content to the portal, which means adding more services and making more departments publish to the new portal rather than the old. These are both major tasks, the former involving some technical challenges, and the latter involving substantial organizational challenges.

References

[Garshol06a] Garshol, L.M., Maicher, L.: Report on the Poster and Open Space Sessions. In: Maicher, L., Sigel, A., Garshol, L.M. (eds.) TMRA 2006. LNCS (LNAI), vol. 4438. Springer, Heidelberg (2007)

[Garshol06b] Garshol, L.M.: Synchronizing Topic Maps with external sources. In: Maicher, L., Sigel, A., Garshol, L.M. (eds.) TMRA 2006. LNCS (LNAI), vol. 4438. Springer, Heidelberg (2007)

[Norge.no] Los: Norge.no web; 2007-12-03, http://www.norge.no/los/

TM*chartis* – A Tool Set for Designing Multiple Problem-Oriented Visualizations for Topic Maps

Hendrik Thomas[1], Rike Brecht[2], Bernd Markscheffel[1],
Stephan Bode[1], and Karsten Spekowius[1]

[1] Institute of Commercial Information Technology
[2] Institute of Media and Communication Science,
Technische Universität Ilmenau,
P.O. Box 100565, 98693 Ilmenau, Germany
{hendrik.thomas,rike.brecht,bernd.markscheffel}@tu-ilmenau.de,
{stephan.bode,karsten.spekowius}@stud.tu-ilmenau.de

1 Introduction

J. L. Borges once wrote about a Chinese emperor, who commanded the creation of an accurate map of China [1]. The resulting map was as detailed as possible but therefore it had to match the size of China. What use would make such a map, especially if you have a narrow navigation question, e.g. where the forbidden city is located?

This legend makes the central problems involved in graphical visualization of Topic Maps quite clear [2,3]. Topic Maps is a powerful knowledge representing paradigm, which formal models or one could say charts a relevant problem domain. This results in a high amount of detailed semantic information like topic names, relations, scope information as well as occurrences. To support navigation and retrieval it is obviously not very helpful to visualize all information modeled in a topic map [4]. The resulting question therefore is: Which part of the information should be displayed and which visualization approach is the most suitable one [5,6]?

2 Graph Based Topic Maps Visualization

Common systems, like the Ontopia's Vizigator [7], are using a topic centered approach, where all associated nodes are displayed in a graph network. This kind of visualization can be applied to any topic map because the graph can be generated automatically using appropriate graph algorithms [6,8]. Despite these clear advantages, this concept possesses three major drawbacks. First, every time a user selects a new topic node, a different graph has to be generated. This results in constantly changing visualizations, which is quite confusing for the user because he can not establish a mental model of the information structure. Second, information on the structure of the semantic information, in terms

L. Maicher and L.M. Garshol (Eds.): TMRA 2007, LNAI 4999, pp. 36–40, 2008.

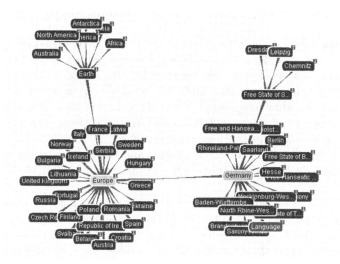

Fig. 1. Example of a graph based Topic Map visualization

of the big picture, are difficult to obtain for the user. The generated graph for a whole semantic network is from the perspective of the algorithm optimal, because the distance between the displayed nodes is maximized. However, from the perspective of the user, the graph is chaotic and difficult to understand due to the ignorance of the meaning of the semantic relations as well as the missing distinction between the ontology and the instance level (see Fig. 1). Third and probably most important, the resulting visualization is dominated by the interpretation and specific modeling decisions of the experts, who created the topic map. However, if Topic Maps are used to support retrieval and navigation, the interpretation of the users can be more important [9]. Based on their individual needs and capabilities only a specific fragment of the semantic net as well as quite different graphical mappings can be appropriate like a hierarchical tree, a list or a network [6,10]. Topic Maps can even be represented as a knowledge map, using the metaphor of a topological map [6]. The visualization choice should therefore depend on the focused task of the user e. g. providing an overview, explaining the knowledge structure or showing detailed information on a field of interests [8]. In the end every information structuring principle or metaphors is suitable as long as it helps the user to understand the modeled semantic information [11]. Practice-oriented domains of application are typically very complex, so often only a combination of different visualization concepts is suitable, e.g. for a digital library [12,13]. A pure automated generation is thereby not recommendable because aesthetic, educational and design aspects can not be considered sufficiently. Additionally, not all required information are modeled – Topic Maps are for modeling knowledge, not for visualization information, e. g. handling of red-green color blindness. In summary, a single topic map visualization is not suitable to satisfy the users' demand for navigation aid.[14].

3 Intelligent Design Topic Maps Visualization

A typical retrieval problem is for example the search for information about the relationships between different concepts, e.g. where is Leipzig located in Europe? In a traditional knowledge base, like an encyclopedia, the user has to search for each concept and the relation separately, e.g. what is Leipzig, which country are part of Europe, etc. How much easier it would be, if someone has copied just the relevant sections out of the encyclopedia and arranged them on a white board in the correct way to help the user understand the relations. Such a problem oriented view would reduce the retrieval time and would prevent the user to be overwhelmed by the huge amount of other information modeled in the knowledge base.

The same stand for Topic Maps, where knowledge of a problem domain is modeled in a holistic way. An automated graph visualization emerge in an evolutionary sense direct from the topic map, e.g. classes are mapped to colors and the position depends on the amount of displayed nodes and edges. This is not sufficient to support a user-specific retrieval. Necessary are problem-oriented views, which focus more on the individual requirements of the user and the retrieval task rather than a generic visualization of the semantic information.

To create such problem-oriented views a human interaction is necessary. Only a human editor is able to interpret a specific problem and process the modeled semantic information in a way to support information retrieval and navigation. Such an intelligent design approach shifts the focus from the automated generation to the design process where manually visualization information are added, e.g. selecting of important nodes, specification of the node arrangements as well as the highlighting of important aspects. To sum up, the human editor must choose which part of the topic map should be display and which structural principles are suitable for the current problem.

4 TM*chartis*

Based on these ideas of a intelligent design approach, we developed the tool set TM*chartis*[1] for designing multiple problem oriented Topic Maps visualizations. TM*chartis* consists of two major applications developed in Java using TM4J, Hypernate and a MySQL database.

First there is the "TM*chartis* editor", a stand-alone java application, enabling experts to design visualizations for individual tasks or navigation problems. Any valid XTM topic map can be imported into the system. Based on an automated initial layout, all topic nodes are positioned on a two dimension drawing map (see Fig. 2). Similar to Google Maps, the user can navigate freely on the drawing map, zoom in as well as search for topic names. The expert can align the topic nodes freely on the map and therefore is able to use any suitable organizational principle in the visualization. The shape and color of the nodes as well as the

---

[1] The name TM*chartis* combines Topic Maps with the Greek word "chartis" means map – in the scene of creating maps for Topic Maps.

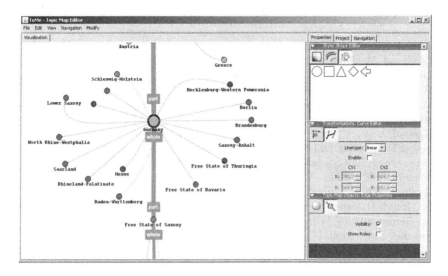

Fig. 2. TM*chartis* Editor

association edges can be customized, for example the path of every individual edge can be linear or curved defined freely. Any non relevant topics, associations or classes can be hidden. All relevant visualization information, e. g. the x and y coordinates of the topics are stored in a central database. For any topic map different visualization can be designed independently, which allows the creation of multiple views on the same semantic information.

The second tool is the "TM*chartis* webviewer" which enables the integration of selected topic map visualizations into any web page as a Java applet. The tool loads the selected topic map as well as the required visualization information form a central database. Within the applet users can move freely on the map, zoom, and search as well as access detailed information about a specified topic, for example all names and occurrence are displayed in a pop-up menu.

5   Summary

Different navigation and retrieval problems require different visualization solutions. Instead of using more and more sophisticate graph algorithms, the answer is the understanding that only a combination of automated layout and intellectual design can provide the user with helpful visualization of the modeled knowledge domain. TM*chartis* addresses these problems by providing an concept for an intelligent design visualization approach and a first prototypical implementation of a tool set for creating multiple views on Topic Maps.

Open tasks in our research are the extension of the editor functionality, like the integration of relevant external information or the development of a semi-automated tool to assist an editor by arranging topics on predefined structuring principles.

References

1. Borges, J.L.: Everything and nothing. New Directions, New York (1988)
2. Le Grand, B., Soto, M.: Visualisation of the Semantic Web: Topic Maps Visualisation. In: IEEE IV 2002, London (2002)
3. Weerdt, D.D., Pinchuk, R., Aked, R.: TopicMaker – An Implementation of a Novel Topic Maps Visualization. In: Maicher, L., Sigel, A., Garshol, L.M. (eds.) TMRA 2006. LNCS (LNAI), vol. 4438, pp. 33–43. Springer, Heidelberg (2007)
4. Garshol, L.M., Moore, G.: ISO/IEC JTC1/SC34, Information Technology – Document Description and Processing Languages (2006),
   http://www.isotopicmaps.org/sam/sam-model/
5. Schumann, H., Müller, W.: Visualisierung – Grundlagen und allgemeine Methoden, Berlin (2000)
6. Le Grand, B., Soto, M.: Topic Maps, RDF Graphs, and Ontologies Visualization. In: Geroimenko, V., Chen, C. (eds.) Visualizing the Semantic Web – XML-Based Internet and Information Visualization. Springer, London (2006)
7. Vizigator Ontopia (2007), http://www.ontopia.net/solutions/vizigator.html
8. Albertoni, R., Bertone, A., De Martino, M.: Semantic Web and Information Visualization. In: Semantic Web Applications and Perspectives (SWAP), 1st Italian Semantic Web Workshop, Ancona, Italy, December 10, 2004 (2004),
   http://semanticweb.deit.univpm.it/swap2004/program.html
9. Peterson, E.: Beneath the Metadata – Some Philosophical Problems with Folksonomy. D-Lib Magazine 12(11) (2006),
   http://www.dlib.org/dlib/november06/peterson/11peterson.html
10. Waniek, J., Brunstein, A., Naumann, A., Krems, J.F.: Interaction between Text Structure Representation and Situation Model in Hypertext Reading (Special Issue: Studying the Internet: A challenge for modern psychology). Swiss Journal of Psychology (2003)
11. Meier, P.: Visualisierung von Kommunikationsstrukturen für kollaboratives Wissensmanagement, Doctoral Thesis, Konstanz (2006),
    http://deposit.d-nb.de/cgi-bin/dokserv?idn=981056008
12. Brix, T., Döring, U., Trott, S., Rike, B., Thomas, H.: The Digital Mechanism and Gear Library: A Modern Knowledge Space. In: Jantke, K.-P., Kreuzberger, G. (eds.) Knowledge Media Technologies – First International Core-to-Core Workshop. Diskussionsbeiträge des IfMK. TU Ilmenau Institut für Medien- und Kommunikationswissenschaft, Castle Dagstuhl, Germany (2006)
13. Thomas, H., Redmann, T., Markscheffel, B.: Controlled semantic tagging - how can topic maps support subject indexing in digital libraries? In: Shoniregun, C.A., Logvynovskiy, A. (eds.) Proceedings of the International Conference on Information Society (i-Society 2007), Merivil, USA, pp. 346–352 (2007)
14. Rasmussen, E.: Information Retrieval Challenges for Digital Libraries. In: Chen, Z., Chen, H., Miao, Q., Fu, Y., Fox, E., Lim, E.-p. (eds.) ICADL 2004. LNCS, vol. 3334, pp. 93–103. Springer, Heidelberg (2004)

Open Educational Topic Maps: A Text-Oriented Perspective

Lars Johnsen

University of Southern Denmark
Engstien 1, 6000 Kolding, Denmark
Larsjo@sitkom.sdu.dk

Abstract. This article makes the case for a text-oriented approach to the creation and presentation of open Topic Maps based information architectures for e-learning. More specifically, it argues that importance should be attached to the communicative aspects of such architectures, in particular consistency, cohesion and coherence. The article further suggests that one way of keeping this aim in focus is to work with standardized text modules as communication vehicles. To exemplify more concretely the role that standardized text might play, two specific sets of tools for modular text production are briefly discussed, namely CNXML and Structured Writing. Finally, it is demonstrated how a descriptive framework such as Rhetorical Structure Theory may be used to support the analysis, design and creation of communicative structures in open educational Topic Maps architectures.

Keywords: Educational Topic Maps, e-learning, CNXML, Structured Writing, Rhetorical Structure Theory.

1 Background

Traditional teaching and learning genres are converging in many subjects and disciplines. Within the field of language learning, for instance, classic teaching and learning resources such as dictionaries, grammars and handbooks today, in varying degrees, contain sections on grammar, lexis (vocabulary) and culture. At the same time, many current teaching materials draw on, or link to, resources available on the World Wide Web. And increasingly, these resources contain open content, that is to say content which may be freely copied, altered and reused at a low, or no, cost.

New types of learning resources call for innovative ways of organizing, integrating and presenting content although traditional information architectures and presentation formats seem to die hard in many products and materials, even electronic ones. As for converging language learning resources, for example, there is an obvious need for the development of information architectures that can accommodate hierarchical text structures on grammar and culture in lexical networks and vice versa.

As has already been convincingly demonstrated in several approaches and projects [1] and [9], Topic Maps provide a very flexible model and technology for organizing and accessing educational content. It is the premise of this article, however, that importance needs to be attached to the communicative aspects of topic maps in order for

L. Maicher and L.M. Garshol (Eds.): TMRA 2007, LNAI 4999, pp. 41–50, 2008.

them to function effectively as (open) e-learning materials. In particular, it is assumed that the educational value of Topic Maps architectures is in no small measure contingent upon the extent to which learners find them consistent, cohesive and coherent.

2   The Seven C's of Educational Topic Map Architectures

Educational topic maps may be said to have three overall goals. Firstly, they must convey information about a certain domain or field (subject matter); secondly, they must do so in a fashion that is lucid and logical for learners (communication); and thirdly they must be modelled and organized in such a way as to be reusable and interoperable (reuse and interoperability).

These goals may be translated into some requirements – the seven C's – that open educational information architectures based on Topic Maps should attempt to meet:

Concepts: Open educational Topic Maps architectures should, if possible, contain conceptual systems that reflect knowledge structures in the chosen domain. The ability to represent conceptual structures is a powerful capability of Topic Maps and one that distinguishes Topic Maps from most other instructional technologies.

Content: Content, often referred to as learning objects, must be included to provide information about subject matter as well as to initiate relevant learning activities (problem solving, exploration, discussion, etc.)

Cohesion & Coherence: Information items must be connected in a manner that supports their communicative goals. Cohesion and coherence are both notions taken from text linguistics and signify ways to achieve unity and flow in spoken or written discourse. While cohesion is about texture in documents and discourse, coherence is (in some theories) about rhetorical organization.

Context: Views and perspectives should be placed on content to indicate pedagogical and didactic contexts (subject, level, learning mode, etc.). In Topic Maps, scope is often the natural way of implementing contexts.

Connectivity: Hooks are needed to link content from dispersed or disparate sources. The whole system of subject identity identification (via subject locators, identifiers and indicators) in the Topic Maps paradigm is designed just for this purpose.

Compliance: Open educational topic maps must conform to technical, legal and ontological guidelines, rules, standards or formats. For instance, topic maps must be represented in a recognized syntax; they must adhere to constraints of underlying ontologies; and to be truly open they must be based on a licence that allows free copying and reuse of content.

The seven C's may be placed in a matrix structure like the one below.

In the matrix the three columns represent the overall goals of an educational Topic Maps architecture. The two rows in the matrix may be said to indicate different angles of looking at the architecture. The first row is essentially about the architecture viewed as an information system ('looking inside') whereas the second row focuses

Table 1. The seven C's of educational Topic Maps architectures

	Subject Matter	**Communication**	**Reuse, interop-erability**
Information system	Content	Cohesion & Coherence	Compliance (technical, legal and ontological)
Domain	Concepts	Contexts	Connectivity

on aspects of its surrounding environment ('looking outside'). Thus, for instance, cohesion and coherence become properties of the internal structure of the architecture. In contrast, concepts signify selected objects in the domain being modelled and contexts designate relevant aspects of the learning environment.

3   The Structure of Open Educational Information Architectures

Information architectures based on Topic Maps are often shown to have two layers: a knowledge layer and a resource layer (often consisting of highly heterogeneous content objects).

In this article an alternative approach is suggested in which an open information architecture for educational purposes contains three layers: a *resource* layer, a *knowledge* layer and a *communication* layer:

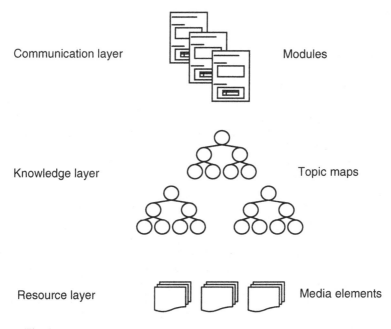

Fig. 1. The three layers of an open information architecture for e-learning

The resource layer contains *media elements*, the basic learning objects that go into a learning architecture (procedures, definitions, pictures, video clips, sound files, etc.)

The knowledge layer holds *topic maps*, representations of interrelated structures in a standards-based model, notably concept structures, content structures, communication structures and contexts.

The communication layer comprises, among other things, *modules*, self-contained text units of predictable size and appearance about one topic or specific aspects of one or more topics. Modules are static, dynamic or layered documents conforming to the same format, standard or set of writing guidelines. They are intended to be consistent, cohesive and coherent presentations, or *re*-presentations (= alternative presentations), of media elements and/or topic maps data.

Modules have two roles: as output in communication situations to convey learning content to learners and as input in (semi-)automated topic map generation processes. In addition to modules, this presentational layer typically includes navigation structures, search facilities and so on.

In this model there is some emphasis on the textual nature of modules in the communication layer. The reason is twofold:

Firstly, it is considered to be crucial that learning content can be presented in meaningful, consistent and measured units that have communicative goals that can be recognized, and recognized consistently, by *learners*. Secondly, practical experience shows that some *authors*, at least in the humanities, are disinclined to think of learning content creation in terms of Topic Maps and Topic Maps based modelling. Writing *text* is their bread and butter, as they see it, not devising clever conceptualizations. Therefore, there has to be some way for authors to contribute to topic map creation through textual means.

4   Creating Modules with CNXML

Luckily, the world of open content does in fact provide us with a set of tools for producing topic-oriented textual content for learning. In the Connexions project at Rice University, open learning content is created, shared and distributed as modules using an XML-based markup language, CNXML, which is freely available from the project website (see http://cnx.org).

CNXML comprises a wide variety of tags to mark up text elements in modules: quite a few structural tags (section, paragraph, link and so on), some semantic ones (example, definition, meaning, etc.) and one or two presentational ones (emphasis). To support authors writing modules in CNXML, the Connexions project offers various tools such as editing and validation facilities, stylesheets, access to publication channels, etc.

Being a document-oriented language, CNXML does not as such include elements for describing and marking up conceptual domain structures. Nor does it contain explicit mechanisms for establishing subject identity. But some kinds of subject identification may, however, be done, albeit indirectly. The reason is that certain markup elements like "term" and "foreign" carry an optional "source" attribute whose semantics allows it to be employed as a sort of subject identifier. If used consistently, this scheme can add some valuable "Topic Maps connectivity" to Connexions modules facilitating harvesting, and subsequent merging, of topics (and related data) from individual CNMXL modules.

CNXML is not the only model for creating modular text in XML, though. A language like DITA is an obvious alternative particularly for training materials or documentation in technical or highly specialized fields [2].

5 Structured Writing

Needless to say, a language like CNXML, or DITA, does not in itself guarantee adequate, let alone user-friendly, communication either. An author may choose to embed all information in a module in massive wall-to-wall text, or he may entirely leave out definitions, explantions and examples in the text – and yet comply with the rules of the CNXML specification.

To enhance the communicative value of CNXML, one may therefore consider applying the guidelines of Structured Writing, a communication methodology devised and developed by R.E. Horn [5], [6] and [7]. Structured Writing consists of guidelines for analyzing, organizing, presenting and sequencing information based on the communicative intent or instructional purpose of the information.

Essential to Structured Writing is the idea that informational content should be divided into manageable units (the principle of "chunking"). Two types of units are recognized in Structured Writing: information blocks and information maps. The smallest unit is the information block, a content chunk with one central informational objective, while information maps are communication modules consisting of a limited number of information blocks, normally five to nine, about the same topic. In terms of size, information maps roughly correspond to one or two printed pages, one web page or one CNXML module.

Information blocks and maps are always labelled to support scanning, skimming or browsing content (the principle of "labeling"). A label may either indicate the contents of an information unit or its communicative function.

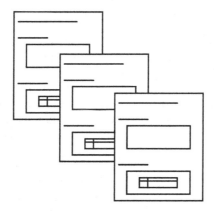

Fig. 2. Information maps

As already mentioned, an information unit should only contain content necessary for the specific purpose at hand (the principle of "relevance"). Content items are classified on the basis of the primary communicative or instructional objective they are designed to fulfil. Seven basic categories, or information types, are recognized:

- Concept (introducing and exemplifying concepts and ideas)
- Procedure (telling the reader how to carry out specific tasks)
- Process (communicating how a series of events develop over time)
- Structure (describing or illustrating objects and their components and possibly the functions of these)
- Fact (providing factual statements such as specifications, results, measurements, etc.)
- Principle (stating policies, rules, axioms and the like)
- Classification (dividing domains into classes, subclasses and instances)

In essence, the principle of relevance ensures that content items belonging to different information types, say a procedure and a definition, are never placed in the same information block.

Another key dictum of Structured Writing is that communicative or instructional goals should be consistently reflected in information or document design, structurally, visually and linguistically (the principle of "consistency"). This means that content items having the same purpose must always follow identical design patterns or templates. Although Structured Writing does not prescribe specific information design patterns, especially tabular formats are widely used in methods based on Structured Writing. This is particularly true of Information Mapping®, a commercially available writing method employing and extending the principles of Structured Writing.

6 Information Maps and Topic Maps

There is a certain conceptual affinity between topic maps and information maps. An information map may be likened to a topic and the information blocks it contains to the occurrences of that topic. And block labels may be perceived as occurrence types specifying relations between topics and resources.

Information maps thus provide us with a kind of basic unified presentation format that may be used for media elements and topic map data alike. A block documenting, say, the structure of a domain object may sit as a formatted file in the resource layer or be the rendition of a set of associations in a topic map. But to the learner the block will look the same. In many, perhaps most, cases an information map will simply be the rendition of a topic and its characteristics. But an information map can in principle present any measured subset of content or data deemed to be relevant for some pedagogical purpose.

Structured Writing also gives authors a simple way of thinking and talking about content layering. Content layering will always consist in adding blocks or maps to blocks or maps in a Lego-like fashion, and it can always be described in terms of laying facts upon facts, providing facts about concepts, attaching concepts to processes, etc.

A consistent, more visually oriented, module structure also facilitates more reliable information extraction from marked up text. In particular, the combination of the semantic tags in CNXML and the tabular formats often found in writing methods based on Structured Writing can be utilized to capture domain knowledge, notably procedures, processes and conceptual structures, and map them onto topic map constructs.

7 Cohesion and Coherence

Cohesion and coherence constitute texture and flow in discourse and documents.

Cohesion is the glue, so to speak, that makes words, sentences and paragraphs connect. Cohesion is created through linguistic markers such as pronouns ("he", "she", "it", etc.) and adverbial phrases ("moreover", "as already mentioned", etc.) as well as through typography and layout, whose main purpose is to affect the reader's perceptual sense of what goes together and what does not.

In a modular approach based on CNXML and/or Structured Writing, linguistic cohesion must be confined to individual modules and must not reach across module boundaries. For example, it makes no sense to refer back to a module with an expression like "as demonstrated above", because there is no knowing whether the reader has actually read the module being referred to.

Linguistic and visual consistency across modules, on the other hand, are essential because it makes it easier for the reader to recognize specific communicative goals or information types. And since consistency is in fact an inherent principle of Structured Writing, it almost comes for free (provided, of course, that the principle is consistently adhered to!).

Coherence is about establishing semantic unity in text. Put somewhat simply, coherence may be construed as the extent to which readers are able to identify the communicative structure that exists between elements in a document, typically spans of text.

Coherence is a central concept in many text linguistic theories and descriptive frameworks, for example Rhetorical Structure Theory (RST) [8]. One of the main tenets of RST is that text often consists of hierarchically organized structures in which individual constituents are related to each other functionally. Constituents may either be nuclei, text elements with a prominent communicative role, or satellites, elements that only play a supportive role.

Nuclei and satellites are linked by so-called rhetorical relations that manifest certain communicative functions. So far twenty odd rhetorical relations have been empirically identified and defined in RST but the class of relations is, in principle, open. The table below lists a couple of common rhetorical relations and briefly explains the semantic characteristics, or roles, of the text constituents that the relations connect.

(Examples taken from the introduction to RST on the RST website: http://www.sfu.ca/rst/01intro/intro.html)

Table 2. Some rhetorical relations and their associated roles

Rhetorical relation	Nucleus	Satellite
Background	Text whose understanding is being facilitated	Text for facilitating understanding
Elaboration	Basic information	Additional information
Preparation	Text to be presented	Text which prepares the reader to expect and interpret the text to be presented

8   Rhetorical Structure in Educational Topic Maps Architectures

With its emphasis on text constituents, relations and roles, RST appears to be a descriptive framework that lends itself to the Topic Maps paradigm. In fact, why not use it as a foundation for analyzing, designing and representing communicative structures among content items in educational Topic Maps architectures?

Most obviously, it will allow us to define rhetorical relations between existing (sub)modules such as information maps and blocks: module B may be described as an elaboration of module A; module C is a summary of modules X, Y and Z and so forth.

But rhetorical organization may also be applied to topic map data directly. One or two examples from the field of grammar teaching may demonstrate this:

The definition of a typing topic may, through reification, be connected to the description of an instance topic to provide background information. For instance, a learner trying to absorb what a "restrictive relative clause" is might be given access to the definitions of superordinate concepts like "relative clause" "dependent clause" or even "clause".

And explaining the difference between, say, the English words "some" and "any" in declarative and interrogative clauses respectively might be done by attaching two textual explanations, in two different scopes ("declarative clause" and "interrogative

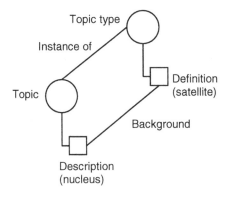

Fig. 3. Rhetorical relation and roles used to signal communicative structure among topic characteristics

clause"), to a topic reifying a rhetorical contrast relation between the two words. The rhetorical relation may here be seen as a kind of predicate allowing us to carry out communicative actions like contrasting, comparing and juxtaposing.

Finally, rhetorical relations may be used to link reified associations between topics. In this way, simple text structures like blocks can be generated directly from topic maps. This, of course, requires some basic data-to-text mapping rules.

Like other conceptual structures, rhetorical structures may be made visible to learners as a kind of communicative scaffolding. For example, one may decide that background information should always be available to users in a separate pane, in a separate window or as a tooltip.

9 Concluding Remarks

Above, it has been suggested that standardized text modules may constitute useful vehicles for consistent and measured communication in Topic Maps based information architectures for e-learning. Admittedly, consistent and measured communication does not in itself ensure *learning* although, no doubt, it is a significant prerequisite. What is needed, therefore, is more work to find out more precisely what types of content combine to form effective topic-oriented learning objects and to examine how these learning objects are actually accessed, used and evaluated by real users in specific learning and teaching environments.

What is also needed to ensure a more widespread use of open educational topic maps is tools to support learning content developers and users such as authors and teachers. Various types of tools may be conceived of. For instance, it would be ideal to have a tool that would allow authors to create, or import, modules in CNXML and organize them, and their contents, in a topic map. This could presumably be achieved by adding functionality to a software package like TM4L [1], which seems to be a very valuable tool for ordinary users who want to be shielded from the most technical details of topic maps. Another step forward would be to set up open repositories to which educational maps could be uploaded, merged and shared. And there is no obvious reason why this could not be done within the framework of the Connexions project thus bringing about tighter integration between open learning content creation and open learning content organization.

References

1. Dicheva, D., Dichev, C.: TM4L: Creating and Browsing Educational Topic Maps. British Journal of Educational Technology - BJET 37(3), 391–404 (2006)
2. DITA, http://dita.xml.org/
3. Johnsen, L.: Designing Adaptive Documentation with XML: From Formal to Rhetorical Markup. Best Practices 5 (June 2003)
4. Johnsen, L.: Information Maps Meet Topic Maps. From Structured Writing to Mergeable Knowledge Webs with XML. In: Madsen, B.N., Thomsen, H.E. (eds.) Terminology and Content Development. Litera (2005)
5. Horn, R.E.: Mapping Hypertext. Analysis, Linkage, and Display of Knowledge for the Next Generation of On-Line Text and Graphics. The Lexington Institute, Lexington (1989)

6. Horn, R.E.: Structured Writing at Twenty-five. Performance and Instruction 32 (1993)
7. Horn, R.E.: Structured Writing as a Paradigm. In: Romiszowski, A., Dills, C. (eds.) Instructional Development: State of the Art, Educational Technology Publications. Englewood Cliffs (1998)
8. Mann, W.C., Thompson, S.A.: Rhetorical Structure Theory: Toward a functional theory of text organization. Text 8(3), 243–281 (1988)
9. Nordeng, T.W., Guescini, R., Karabeg, D.: Topic Maps for polyscopic structuring of information. Int. J. Continuing Engineering Education and Lifelong Learning 16(1/2), 35–49 (2006)
10. Park, J., Hunting, S.: XML Topic Maps. Creating and Using Topic Maps for the Web. Addison-Wesley, Boston (2003)

Using Topic Maps for Visually Exploring Various Data Sources in a Web-Based Environment

Eicke Godehardt[1,2] and Nadeem Bhatti[2]

[1] SAP AG, Germany
eicke.godehardt@sap.com
[2] Fraunhofer IGD, Germany
{eicke.godehardt,nadeem.bhatti}@igd.fraunhofer.de

Abstract. Today every company and every single person owns a lot of data. However, easy ways for exploration and navigation are not very widespread. Especially, if complex relations between these data are defined and the amount of data becomes bigger, an easy to use interface is necessary.

This paper describes a topic map viewer, which is easy embeddable into websites to offer none-experts an innovative way to find new and relevant information. To display even huge information spaces in a web environment, this paper describes two concepts supporting scalability are explained, namely data exchange and visualization. For the data exchange between client and server pre-fetching mechanism is introduced to load data in small chunks. Garbage collection complements this to prevent overloading of the viewer by unloading not needed topics. Furthermore the clustering concept for the visualization is used to handle the huge amount of information.

1 Introduction

Most of the data sources today are not accessible at all or only understandable by experts. Especially if complex relations between these data are defined and the number of data becomes bigger, an easy to use interface is necessary to allow also non-experts to browse through this data and gain new information.

In this paper, we describe how we use XML Topic Maps to visualize data in a web-based environment. Topic maps provides a model and grammar for representing the structure of information resources [1]. The topic map viewer displays the topic map and allows the users to navigate and explore through the map.

The goals and requirements for the topic map viewer are the following:

- web-based and easy integration in existing websites (Web2.0)
- visualizing already existing data sources
- large amount of data
- easy navigating/exploring through the data

Based on the requirements and goals above, the following key points will be explained. How we use existing databases via a XML Topic Map interface? In addition we focus especially on using different data sources and how we solve the issue of handling very large data sets.

L. Maicher and L.M. Garshol (Eds.): TMRA 2007, LNAI 4999, pp. 51–56, 2008.
© Springer-Verlag Berlin Heidelberg 2008

The paper is organized as follows. Section 2 gives a brief overview on related work, which are using topic maps as a basis for some kind of visualization. Section 3 describes our approach, in particular how we retrieve the data as XML Topic Maps and handle scalability issues through loading on demand and pre-fetching. Subsequent in section 4 we show a use cases for this technology. We summarize our findings in section 5 and give an outlook on what to solve in future increments of the topic map viewer.

2   Related Work

There are several approaches for using topic maps for visualization in a web based environment. Below, some of these works will be mentioned and described in brief.

TopiMaker[2] is a stand alone application, which has the main goal to provide a novel authoring environment. It positions the topic maps into several layers placed in a 3D environment. The visualization of TopiMaker is more suitable for authoring and exploration is not widely supported. The second main difference is that TopiMaker is designed to be a stand alone application in contrast to a web application as our solution is designed for. Another examples which focus more on editing are [3] and [4], but the described solutions also offers only a very primitive view, which is not sufficient for exploring huge and complex data sources.

There is also some research going on about visualization and navigation related to our work. For example a simple topic map viewer [5] or visualizing auto-generated topic maps [6]. In addition there is also some work in visualizing search results or metadata [7]. The systems are mostly Java based and desktop applications.

While there are several approaches and some of them consider related goals none of them focus on scalability or handling large topic maps in an embeddable web-based environment.

3   Approach

In this section the main way of acquiring the data is described. Figure 1 below gives an overview of the data flows on how our topic map viewer accesses different data sources. The response is encoded as an XTM – XML Topic Map (XTM [1]) file.

3.1   Scalability

The goal of our visualization tool was to allow an easy integration into existing websites to visualize existing data. Options for the visualization technology in web are Flash and using Java Applets. Graphical visualization is not possible with Javascript alone and SVG is not yet widespread. But for Applets it can become very complex to get them running in every single environment. To reach the main objective of this paper, the Flash technology and XML Topic Map format as the data exchange format are used to achieve a web-based solution. The Flash Player is a client application and available in most common web browsers (over 98% [8]) and supports vector and raster graphics and an object-oriented language called Actionscript (AS).

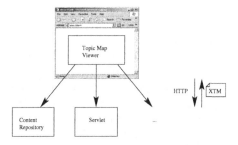

Fig. 1. General access of data sources

On the other hand, this decision, which satisfies the required flexibility, has an important drawback when it comes to scalability. A straight forward realization may handle a few hundred topics. But in real-word scenarios this is not sufficient. In large companies and even small enterprises the number of topics easily becomes several thousand. But as it is not the goal to visualize all topics at once. Instead the topic map viewer will support kind of exploration behavior by loading the data on demand. The user chooses one starting point in the topic map. Then the topic map viewer displays a small amount of information around this starting point (chosen topic). Afterwards s/he can navigate through the map by selecting different topics according to their own interest. By selecting the viewer requests for small packets or chunks from the server (described above), which allows an interactive and responsive user interface. Every response from the server is still a complete XTM file (see figure 1) and is merged with current state of topic map inside the viewer.

The user may not even know that not all data is loaded at the beginning of his/her survey. The figure 2 below points out this behavior. It is very important to notice that the figure 2 shows the status of the loaded topic map and *not* the visualization of the topic map itself.

Fig. 2. Loading parts of the topic map on demand

In the image three steps of changing the current topic is shown and how the internal topic map is changed accordingly. The topic maps are loaded on demand as the user clicks on a different topic and merged with the already load topic map inside the viewer. This is only done once fore every new topic to reduce traffic, as we do not expect the topic map to change at the server side.

To merge all responses from the server and integrate more and newer information, the topic map viewer technically uses the original definition of mergeMap. mergeMap is defined by the TopicMaps.Org Authoring Group for XTM (see [9]) and normally allows the subdivision of a topic map over more than one file.

3.2   Pre-fetching

The above described technique can be improved as there is still some latency when the user selects a new topic, as the new data chunk has to be loaded. We achieve this by pre-fetching or pre-loading additional parts of the topic map in advance. For example starting from the current topic, the topic map viewer first loads all direct connected topics together with the current one. This is exactly the same behavior as described in the section above. After all direct connected topics are loaded and displayed, the viewer starts loading all topics that are "two steps away" from the current topic, i.e., direct connect topics from all loaded topics in the first place. This can be repeated several times to pre-load the context of the user in advance, e.g., three level context of the current topic as shown in figure 3. When the user starts navigating around in his/her current context, the topics can be displayed immediately as they are already loaded. Additionally pre-fetching starts again in respect to the new current topic.

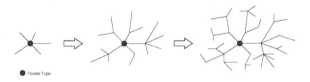

Fig. 3. Pre-loading topics of the current context of the user

As in figure 2 the displayed topic map represents the loaded topics and not the visualization of the topic map, which is subject of section 4 below. Unlike the earlier figure 2, the current topic does not change in figure 3 although the internal topic map inside the viewer is enhanced over time.

The main advantage of pre-fetching the user's context is a smoother user experience. Especially over a slow Internet connection, the general response time while browsing the topic map is much shorter than using the elementary on-demand solution alone described in section 3.1 above. The combination of loading on demand and pre-fetching is still scalable, which means the viewer can handle topic maps of huge sizes while still providing a responsive user interface.

3.3   Garbage Collection

A drawback of pre-fetching and loading on demand is the growth of the topic map inside the viewer. The number of already loaded topics can become really big while using the system. This growth can lead to a slow down of the topic map viewer response time, as more topics have to be handled. To circumvent such a slow down, we suggest kind of a garbage collector to reduce the maximum number of loaded topics. In contrast to the usual definition of a garbage collector, we just focus on the task in freeing memory of our application. While the deleted topics are not garbage in the usual sense (not referenced/accessible), they may not be used in the future any more. But even in this case, they can be loaded once again without problems at any time in the future.

Every topic, which may not be interesting to the user, is dropped. To figure out if a topic is still of interest to the user, the viewer uses some heuristics. The most practical and obvious indicators for such an action could be how long ago is the last time of display are or how far away is a topic from current topic? One possibility could be to delete every topic from the current topic map which was last displayed at least 15 minutes ago or the distance from the currently selected topic is greater than 10 steps. All this may indicate topics, which may not be used in the near future and can thus be deleted to optimize the runtime behavior of the topic map viewer.

4   Use Case/Case Study

The objective of the project "SAP-TM-Viewer" is to develop an SAP applications integrated information system in order to support the knowledge engineers within the enterprise processes. When the knowledge engineers need help during the enterprise process, this information system should provide the context specific help.

In order to achieve the goals of this project, the key approach was to develop a Knowledge map so called SAP-TM-Viewer on the base of the TM-Viewer technology (see above) to visualize the knowledge items, the relationship and hierarchical structure between the knowledge items. The SAP-TM-Viewer consists of fields which are abstracted from the SAP topic map. Knowledge concepts in each field are represented with specific icons, lines between the knowledge concepts represent the associations and the levels represent the abstraction level of the knowledge concept (inner level show generic knowledge concepts) as shown in figure 4.

The pre-fetching, garbage collection mechanisms of TM-Viewer techniques and cluster concept make it possible to handle a huge amount of knowledge items ($¿4000$) while still being responsive. But both concepts do not solve the problem to visualize the huge amount of data. The visualization of huge number of knowledge items, e.g., more than 100 topics can lead to cognitive overload. That is why the cluster concept was added to keep the visualization manageable for the knowledge workers. According to the cluster concept all the topics, which have same sibling, will be clustered as it is shown in the figure below.

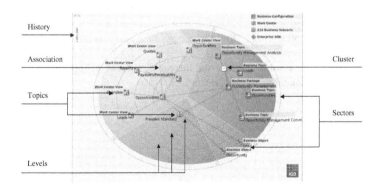

Fig. 4. TM-Viewer

5 Conclusion and Future Work

In this paper we have shown, how we used Topic Maps as a generic kind of data exchange used to view and explore huge data sources in a web-based environment. In particular it was shown, how issues like scalability and responsiveness are handled using loading on demand, intelligent pre-fetching and garbage collection.

Future research directions on the data retrieval part may be performance evaluation by user tests and awareness. Awareness (first addressed in section 3.1) concept will enable the user to get an instant visual feedback of dynamical changes in the topic map. For example, if the enterprise applications change the semantic information the topic map viewer shows these changes in semantic visualization. This approach will keep the visualization up-to-date. Another important focus for future research could be to expand the ability of topic map viewer to visualize other different types of data structures as intuitive as the tree-like structure described above (e.g. process visualization).

References

1. TopicMaps. ISO/IEC 13250.
2. De Weerdt, D., Pinchuk, R., Aked, R., de Orus, J.-J., Fontaine, B.: Topimaker - An implementation of a novel topic maps visualization. In: Proceedings of International Conferences on Topic Maps Research and Applications, Leipzig (2006)
3. Ditcheva, B., Dicheva, D.: Visual Browsing and Editing of Topic Map-based Learning Repositories. In: Maicher, L., Sigel, A., Garshol, L.M. (eds.) TMRA 2006. LNCS (LNAI), vol. 4438, p. 2006. Springer, Heidelberg (2007)
4. Hornung, J., Hornung, C., Oliveira, A., Fernandes Marcos, A.: Knowledge Visualization based on Topic Maps for Computer-Based Learning Applications. In: World Conference on Open Learning and Distance Education (2001)
5. Ontologies for Education Group. Simple topic map viewer: Very general and java applet based o4e (ontologies for education) portal. http://iiscs.wssu.edu/o4e/
6. Amende, N., Groschupf, S.: Visualizing an Auto-Generated Topic Map
7. Howarth, L.C., Miller, T.: Visualizing search results from metadata-enabled repositories in cultural domains. In: Maicher, L., Park, J. (eds.) TMRA 2005. LNCS (LNAI), vol. 3873, pp. 263–270. Springer, Heidelberg (2006)
8. Adobe. Flash player penetration. http://www.adobe.com/products/player_census/flashplayer/
9. TopicMaps. Org Authoring Group. Xml topic map - mergemap specification, http://www.topicmaps.org
10. TM4L. Darina dicheva, towards reusable and shareable courseware: Topic maps-based digital libraries, http://compsci.wssu.edu/iis/nsdl/

Topincs Wiki –
A Topic Maps Powered Wiki

Robert Cerny

An der Embsmühle 25, D-65817 Eppstein, Germany
robert@cerny-online.com
http://www.cerny-online.com

Abstract. Topincs provides a RESTful web service interface for retrieval and manipulation of topic maps. A Topincs Server implementing the interface can host many stores, which are collections of topic maps. The Topincs Editor is a browser-based application that allows editing of topic maps. In this paper, the Topincs Wiki, another view on the content of a Topincs Store, is presented. Its purpose is to hide the Topic Maps paradigm from the user, thus simplifying the collaborative creation of topic maps. The Wiki still emphasizes formal statements. It uses *ontological reflection* to guide users to data conformity rather than enforcing it with a schema.

1 Introduction

With the advent of the Internet in the early nineties of the last millennium, Ward Cunningham took the next step to simplify the creation of web content. He designed and implemented a system for authoring and linking web pages using a web browser only. He describes a Wiki to be "the simplest online database that could possibly work"[5].

Since 1995, users of the original Wiki, the *Portland Pattern Repository's Wiki*[1], are discussing the topics *people, projects and patterns* in the area of software development [4]. Their collaborative effort produced tens of thousands of pages [5]. Since this content repository has been first used, the Wiki idea has found wide acceptance, first within the technical communities and then within the general population. Wikipedia[2], the free encyclopedia, currently contains over 2 million pages [6] and is arguably the most accessed source of facts today.

The essence of the Wiki idea has been summarized on many occasions. For the sake of the argument of this paper, we are condensing the Wiki concept as follows:

– Wiki users author pages by making *statements* about the subject of the page in *natural language*, supported by a simple markup language.

---

[1] http://c2.com/cgi/wiki
[2] http://en.wikipedia.org

L. Maicher and L.M. Garshol (Eds.): TMRA 2007, LNAI 4999, pp. 57–65, 2008.

- Some of the statements on a page refer to subjects that are discussed on other pages. Users have the option of connecting pages with *untyped* and *unidirectional* links.
- Wiki users discuss the statements regarding their truth value, appropriateness and style.
- The Wiki software keeps track of the changes the users make.

1.1 Semantic Wikis

The use of *natural language* makes it easy for users to contribute to the knowledge sharing effort in a Wiki. The use of natural language, on the other hand, restricts the possible knowledge providers and consumers to people who speak the language.

Hyperlinks allow navigation between the pages of a Wiki. These links are *untyped* and *unidirectional*. The nature of the relationship between two pages is only accessible by a person who reads and understands the statement in which the link is rendered. Because the association between pages is unidirectional, the link target is not connected to the source page, leaving the reader of the target page unaware of its relationship to the source page[3].

By using a knowledge representation formalism it is possible to overcome these weaknesses. A system that does not encode information primarily in natural language gains the following features:

- Humans and machines can act as content providers and consumers.
- Connections between subjects are *typed* and *multidirectional*.
- Less effort is needed to translate the content to different languages.

A Wiki system that supports encoding of information in a formal way is referred to as a *Semantic Wiki*. Most existing Semantic Wikis use RDF and related technologies to achieve machine processability. Some solutions in this area are Semantic MediaWiki [9], Platypus Wiki [8] and IkeWiki [7].

This paper presents a solution based on Topincs, which uses the Topic Maps Data Model [3] to encode knowledge. Another project that uses Topic Maps technology in the same area is the Ceryle Wiki[4]. Benjamin Bock and Lars Marius Garshol presented their two separate Wiki approaches, based on Topic Maps, in the Open Space Session of the TMRA 2007 [10]. Tobias Redman and Hendrik Thomas laid out their ideas for the design principles of a Topic Maps based Wiki [12].

2 The Topincs Wiki

The *Topincs Wiki* is part of a larger software system called *Topincs*[5][1]. It consists of a Web Service Interface, a Server, and different clients. At present there

---

[3] Most Wiki systems offer a *What links to this page* function.

[4] http://www.altheim.com/ceryle/

[5] http://www.cerny-online.com/topincs/

are the Editor and the Wiki. One Topincs installation can host many stores, which are collections of topic maps, that expose their items through URLs. The *Topincs Interface* defines the meaning of the application of the HTTP verbs POST, GET, PUT and DELETE on the URLs (item identifiers) that a store exposes. The *Topincs Server* implements the Interface. The current Topincs implementation uses PHP for request processing and MySQL for persistence. It supports JSON Topic Maps (JTM) on writing requests and JTM, XTM 1.0 and XTM 2.0 on reading requests. The *Topincs Editor* is a browser-based user interface for editing topic maps. It is an Ajax application and allows editing of all features of the TMDM [3] except variants and reification.

The *Topincs Wiki* offers another user interface for editing the content of a Topincs Store. It completely hides the Topic Maps paradigm from the user. Whereas other Semantic Wikis offer the creation of formal statements as an add-on to the article writing in natural language, the Topincs Wiki puts formal statements in the center of attention and tries to simplify the creation of these as much as possible. The user is completely unaware that he is using a knowledge representation formalism.

In comparison to the Editor, the Wiki consists of web pages, which are supported in their functionality by Ajax. However, it is not implemented as an Ajax application, because a web browser offers a lot of navigational infrastructure which is hard to incorporate in an Ajax application that loads only once and remains on the same URL during its entire lifetime.

Currently the Wiki allows viewing and editing of articles, but lacks support for discussion and versioning. An implementation of these important features is planned.

2.1 Viewing an Article

In Figure 1, an article about the TMRA 2007 is shown[6]. The content of the page is easily understandable. The reader does not notice a distinction between different statement types. The occurrences *Homepage*, *Start date*, and *End date* look similar to the associations *Organized by*, *Chaired by*, *Takes place at*, and *Consists of*. The players of the associations are rendered as links. By following a link the user navigates to the Wiki page of the player. The locale dependent information is displayed according to the settings of the web browser[7].

Every article is based on a topic map. For latency reasons, all information is delivered within the first page load. This *virtual topic map* integrates information of all *physical topic maps* in the store. It consists of

– the main topic with all names and occurrences,
– all associations that have the main topic as a player, and
– all typing topics and players with minimal information for display.

---

[6] This article is available at `http://www.topincs.com/tmra/2007/wiki/id:142`
[7] The implementation supports currently only the German and US date format.

Fig. 1. The Wiki displays a topic map as a list of occurrences, names, and associations. Players are rendered as hyperlinks.

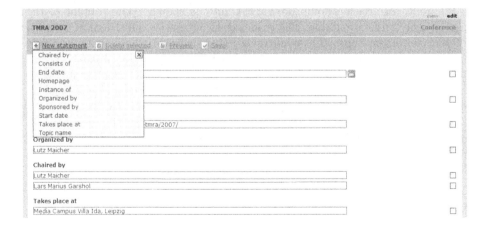

Fig. 2. When editing an article, possible options for new statements are listed using ontological reflection

2.2   Editing an Article

Since our goal is to hide the Topic Maps paradigm from the user, we need to make it transparent whether an occurrence, an association, or a name is edited. We achieve this by simply offering him all choices for new statements about the subject of the page in one control as shown in Figure 2. The Topincs Interface allows the querying of possible statement types for a topic based on its type. Which algorithm is used to create the list of possible statements is up to the server. The current implementation uses *ontological reflection*. If there exists a *Conference* with a *Homepage* within the system, it will offer this statement type when editing any *Conference*. At a later stage it is possible to base the responses to these queries on a schema.

2.3   Editing Occurrences

Occurrences and names are statements within the TMDM, which deal only with one topic. They are attached directly to the topic and connect it to some data, which is represented as a string. Every occurrence has a datatype. The TMDM specifies a set of standard datatypes, which are the XML Schema datatypes [3]. For some of them Topincs Wiki uses specialized input elements. For example, a control for editing dates is shown in Figure 3. In addition to the standard datatypes, Topincs uses the datatype `http://tmwiki.org/psi/wiki-markup` to allow the usage of Wiki markup in occurrences.

Fig. 3. Editing an occurrence of datatype date

2.4   Editing Associations

Creating a user-friendly control for editing associations is in our case particularly difficult. Since Topincs does not use a schema, the set of topics to choose from consists of all topics in the store. Furthermore, the user should be able to create a new topic on the fly. Since assigning a type is important, the control needs an additional input element, which appears only if the user refers to a new topic. This control allows him to specify a topic type, which again can be new.

Any input element in our association control will expect the user to start typing. Once the user pauses, the input element will query the Topincs Server for topics with a matching name. The result will be displayed in a drop down box. If no topic matches, the control switches to the *New topic* mode as shown in Figure 4.

Fig. 4. Editing an association and creating a new topic on the fly

2.5 Saving Changes

Once the user decides to save his changes, the JavaScript program splits up the changes into atomic operations according to the Topincs Web Service Interface [1]. Additions result in a POST request to the topic in case of occurrences and names or to the topic map of the topic in case of associations. Deletions will be DELETE requests to the item identifier of the deleted item, and modifications will PUT a new representation of the item to its item identifier. The user is informed of the saving progress by a bar in the lower right hand of the page.

3 Modelling

Modelling knowledge in Topincs is a continuous process. Topincs was originally designed to support personal knowledge management [1,11]. A human being encounters information from various domains in his everyday life. There are no strict boundaries on the conceptual level, therefore we chose an approach which we call *ontological reflection*. For a topic of type *Person*, statement types are suggested that are already in the store for topics of that type. In order to extend the options for new statements in the Wiki, one has to use the Editor to model the first statement after the creation of the necessary types. From then on the statement type will be offered in the *New statement* menu in the Wiki.

If we want to state that *Thomas Vinterberg is the director of the movie "Dear Wendy"*, we need to create an association type *Director/Directed thing* and two role types *Director* and *Directed thing*. The role names do not matter. We need to name the association type from the perspective of both roles. Therefore we create the name *Director of* in the scope of *Director*, and the name *Directed by* in the scope of *Directed thing*. After this we can create the statement we want to express. All this has to be done in the Topincs Editor. From now on, all statements of the same form can be created in the Wiki.

With this simple methodology it is possible to offer the user some guidance about what statements can be expressed about a topic of a certain type. *Ontological reflection* could as well be used to fill drop down boxes with candidates for a role. Yet, it is not possible to model restrictions on cardinality, e.g. a *Person* can have three birth dates and two mothers. We have been reluctant in implementing a schema language in Topincs because we take the position that we are not modelling reality, but rather what people say about reality. Furthermore the need for conformity of data only arises with computation. Therefore programs should check the validity of their input. Human consumers can, depending on the area of application, deal with missing and contradictory information. For people *trust* is more important than *logic*. The reification mechanism of Topic Maps supports us in keeping track of our encounters with a statement, but also allows us to express statements like *Harry believes that John is born in 1975*. Therefore the integration of reification in the user interfaces of Topincs will be introduced in one of the next versions.

4 Language Independence

Information in a topic map is stored independent of any natural language. Together with the scoping mechanism of Topic Maps, it is possible to present information in different languages. As soon as the ontology is translated, the information of a topic map is intelligible to speakers of that language, because names of individuals do not need translation in most cases. In our context, an ontology is a set of topic types, occurrence types, name types, association types and role types [2].

To use this feature of Topic Maps, Topincs will receive a view to translate names from one language to another. The view will work on a customizable set of topics. The Topincs Server will choose the display names according to the language settings of the web browser. It might be useful to allow the user to override that setting within the web page. It is important not to break the caching mechanism of the HTTP protocol when serving items in a specific language. It must be clear that a proxy cannot satisfy a request, if the locale information does not match. Another option would be to deliver the topic map that makes up an article and fetch the display names in separate requests. Even though we like this idea conceptually better, we did not implement it, because of the impact on the usability of the Wiki. Since the information can only be displayed with names, the page rendering would have to be stalled until all names have arrived. Since browsers usually process only a few requests simultaneously, it might be very unpleasant for the user, since there are many names to be fetched.

With this solution in place, it will be possible for people to collaboratively create information over the world wide web across language boundaries.

5 Application of Topincs at TMRA 2006 and 2007

In 2006 and 2007, Topincs was used to record knowledge about the TMRA conference, its presentations and the content thereof [8]. This initiative was driven by the conference organizer Lutz Maicher and the author. Both years participants were encouraged before the keynote to actively take part in this documentation effort.

Tables 1 and 2 show the number of items created during the two conference days in 2006 and 2007. This information was pulled out of the database of the Topincs Stores for the two years by counting the items that were created during

---

[8] Available at `http://www.topincs.com/tmra/200(6|7)/wiki/`

Table 1. Number of items created during TMRA 2006

	Topics	Occurrences	Associations	Total
Day 1	89	42	80	211
Day 2	19	4	33	56
Total	108	46	113	267

Table 2. Number of items created during TMRA 2007

	Topics	Occurrences	Associations	Total
Day 1	57	21	113	191
Day 2	32	8	67	107
Total	89	29	180	298

the two day conference. A comparison is somewhat difficult. In 2006, the Editor was used, whereas in 2007, the Wiki was used. The Wiki creates more statements automatically. For the author, who was also a main content contributor, network connectivity was worse in 2007.

6 Summary and Plans

The Topincs Wiki is a semantic Wiki based on Topic Maps. It emphasizes formal statements, because they are machine processable and language independent. It uses *ontological reflection* to guide users in their knowledge sharing endeavour.

The Topincs Wiki needs a page to discuss a topic and the statements about it. In order to support versioning the Topincs Interface has to be extended to allow retrieval of specific versions. The persistence layer of the server has to archive the current version of an item before replacing it on PUT requests. DELETE requests have to take into account that an older version might be restored.

Topincs proves itself as an invaluable tool to the author. An issue tracking system[9] was set up within a few hours and is extended on demand. At the same time the need for queries becomes evident, e.g. to display all unsolved issues. Therefore we are considering the integration of a query language into Topincs.

References

1. Cerny, R.: Topincs - A RESTful Web Service Interface For Topic Maps. In: Maicher, L., Sigel, A., Garshol, L.M. (eds.) TMRA 2006. LNCS (LNAI), vol. 4438. Springer, Heidelberg (2007)
2. Garshol, L.M.: Towards a Methodology for Developing Topic Maps Ontologies. In: Maicher, L., Sigel, A., Garshol, L.M. (eds.) TMRA 2006. LNCS (LNAI), vol. 4438. Springer, Heidelberg (2007)

---

[9] http://www.topincs.com/issues/wiki/

3. ISO/IEC FDIS 13250-2: Topic Maps — Data Model, 2005-12-16, International Organization for Standardization, Geneva, Switzerland (2005), http://www.isotopicmaps.org/sam/sam-model/2005-12-16/
4. Portland Pattern Repository's Wiki: Welcome Visitior. (Accessed on 2007-11-03). http://c2.com/cgi/wiki?WelcomeVisitors
5. Wiki: What Is Wiki. (Accessed on 2007-11-03). http://www.wiki.org/wiki.cgi?WhatIsWiki
6. Wikipedia, Statistics. (Accessed on 2007-11-03). http://en.wikipedia.org/wiki/Special:Statistics
7. Schaffert, S.: IkeWiki: A Semantic Wiki for Collaborative Knowledge Management. In: Proceedings of the 15th IEEE International Workshops on Enabling Technologies (WETICE 2006), Manchester, U.K (2006)
8. Campanini, S.E., Castagna, P., Tazzoli, R.: Platypus wiki: A semantic wiki wiki web. In: Proceedings of the First Italian Semantic Web Workshop (SWAP 2004), Ancona, Italy (2004)
9. Krötzsch, M., Vrandecic, D., Völkel, M.: Semantic MediaWiki. In: Cruz, I., Decker, S., Allemang, D., Preist, C., Schwabe, D., Mika, P., Uschold, M., Aroyo, L.M. (eds.) ISWC 2006. LNCS (LNAI), vol. 4273. Springer, Heidelberg (2006)
10. Garshol, L.M., Maicher, L.: Report from the Open Space and Poster Sessions. In: Maicher, L., Garshol, L.M. (eds.) Scaling Topic Maps. LNCS (LNAI), vol. 4999. pp. 1–13. Springer, Heidelberg (2008)
11. Sigel, A.: Report on the Open Space Sessions. In: Maicher, L., Park, J. (eds.) TMRA 2005. LNCS (LNAI), vol. 3873, Springer, Heidelberg (2006)
12. Redmann, T., Thomas, H.: The Wiki Way of Knowledge Management with Topic Maps. In: Shoniregun, C.A., Logvynovskiy, A. (eds.) Proceedings of the International Conference on Information Society (i-Society 2007), Merrillville, USA (2007)

Bookmap –
A Topic Maps Based Web Application
for Organising Bookmarks

Tobias Hofmann and Martin Pradella

Bauhaus University Weimar, Faculty of Media, Bauhausstr. 11,
99423 Weimar, Germany
`tobias.hofmann@medien.uni-weimar.de`,
`martin.pradella@com2-gmbh.de`
`http://www.uni-weimar.de/medien/cg`

Abstract. This paper proposes a basic ontology for use in topic maps storing semantic information on bookmark collections. Furthermore, we introduce a data model allowing to implement such a system on a LAMP (Linux, Apache, MySQL, PHP) platform, extended with the Cake-PHP framework. A prototype has been developed as proof of concept, where the use of AJAX and drag and drop capabilities in the browser resulted in a good user experience during a preliminary user evaluation.

1 Introduction

As part of the work for a diploma thesis, we needed to have a bookmark organising system which is server-based and accessible from everywhere. In this system, we wanted to be able to have a hierarchical representation, like we know it from current browser and bookmark systems: trees with folders, subfolders, and bookmark items. Such bookmarks should be imported from existing collections of bookmarks, but also be added at a later date, in a very simple manner, so the system can be used by inexperienced users.

Furthermore, we wanted to be able to connect folders and single items using semantic associations. This is why we decided to include Topic Maps into the system design, which ended up as a web application, due to the internet nature of bookmarks. As a result, all user interaction was to be done whithin a browser window, accessing the server from everywhere. Finally, bookmarks were to be treated differently, depending on their privacy status, and possibly reused by other users of the same system. The name of this system was decided to be bookmap.

In this paper, we will now describe the requirements, some of the design decisions, and architectural elements of the bookmap software.

2 Previous and Related Work

Passin describes in [1] a Topic Maps based system to organise bookmarks, and implements such a system using the TM4Jscript library [2]. Here, the data model

L. Maicher and L.M. Garshol (Eds.): TMRA 2007, LNAI 4999, pp. 66–73, 2008.

focuses on the representation of hierarchies, and integrates Passin's own concept of 'And-Terms' to allow for cross references. As we did not intend to use the hierarchic tree as a central element, but saw it more of a concession to users' current understanding of classifying, Passin's model did not come to use.

Then we have, of course, social bookmarking portals, enabling users to have a central bookmarking organisation. Del.icio.us as one of the first and surely coining representative of its kind has, as Matt Biddulph states [3], three axes: User, Tag, and URI. The users make up the social element, building an information sharing community. The Tags allow the users to classify information differently from, say, taxonomies, in a seemingly more convenient and surely more popular manner. Finally, the URI as the central element of organisation comes in.

Recently, a site with the name Fuzzzy.com has come to life, which came to our attention when development work on the current state of the bookmap system had been finished. Fuzzzy builds an ontology by using tags as topics, and by giving the user the opportunity to create horizontal and vertical associations between those tags. Later, the bookmarks are tagged and thus attached to the ontology.

This is different from our approach: We have a bookmark, which is represented by a topic, a basename and the topic's subject identity (URI). The basename is the topic's and thus the bookmark's name. The subject identity of this topic is the URI. Other users may have their own bookmark, of the same URI, but with another basename. This allows a very individual organisation of private or partly private bookmarks. Furthermore, Fuzzzy seems to suggest to the user to work without a hierarchical representation many users have come to rely on.

3   Motivation and Goals

One of the objectives of the diploma thesis was to identify a scenario in which Topic Maps would be put to good use. The developed software should itself be and rely on open source software. Eventually, a prototype was to be built and evaluated, time permitting, against the requirements collected around a user centered approach.

Technically, and after quite some time, it was decided to go with a LAMP based backend, also due to the high degree of availability in the hosting market. On top of this, and to provide for a flexible architecture with clear structures, we would use the MVC (Model-View-Controller) paradigma, to separate the three layers storage, system logic and presentation, by using the CakePHP framework [4]. The data and logic provided by the backend was to be presented to the user in a browser window, on as many OS platforms as possible, and adhering to W3C recommendations, which led us to Mozilla/Firefox as the browser of choice. Eventually, we would want to have a good desktop integration and high ease of use by permitting the user to drag and drop items, also from the desktop to the browser window, in our aim to meet the user's expectations.

4   Ontology

We want to represent bookmarks in a topic map. To do so, and as mentioned above, we use for each user and each URI a topic of type item which we call bookmark. This bookmark has a basename, and a subject identifier with the bookmark's URI. Fig. 1 shows in more detail how this is designed.

Such bookmarks or items can be connected using semantic associations between two bookmarks. The ontology provides a set of predefined association types (see (Table 1)) including role types, which we found to be sufficient to express the vast majority of relations our users wanted to express. You will find two types of associations, directed, with different role types, and inderected, with only one role type. Furthermore, associations in bookmap are always 1:1, never 1:n.

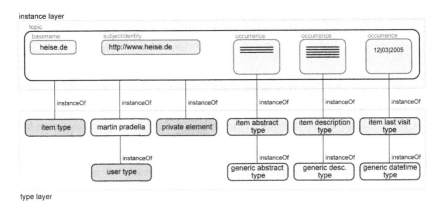

Fig. 1. Visualisation of the data structure of an item-topic

Table 1. Associations provided by the Ontology, including role types

AssociationType	Roletype A	Roletype B
explanation	explains	is explained by
description	describes	is described by
interpretation	interpretes	is interpreted by
demonstration	demonstrates	is demonstrated by
discussion	discusses	is discussed by
recommendation	recommends	is recommended by
summary	summarizes	is summarized by
sequence	precedes	to follows to
part-whole-relationship	consists of	is part of
weblink	links	is linked by
tagging	tags	is tagged by
confirmation	confirms	confirms
contradiction	contradicts	contradicts
relation	is related to	is related to
equality	is equal to	is equal to

There is no constraining in the type of topics allowed to play a role in associations. In this first stage of the software, we wanted to provide a maximum of flexibility and find out how users will use the associations, and felt that limiting their choices of usage at that time would anticipate later design decisions.

5   Database Model

The design of the database schema is of great importance for the prototype, as its mapping of the Topic Maps model has to support the structural flexibility of the Topic Maps paradigma to the full extent. This is why following the proposition of Mugnaini [5], we extended his database scheme to our needs.

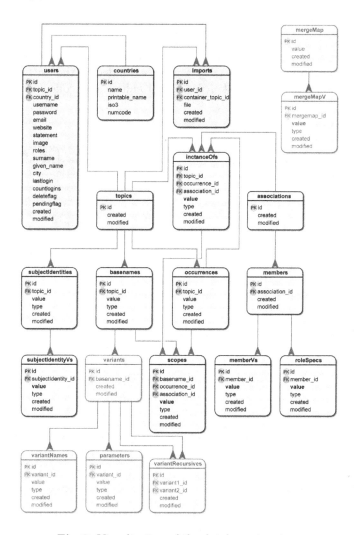

Fig. 2. Visualisation of the database structure

Those extensions mainly concern the ability to later achieve a separation between the topic map containing associations between bookmarks and system data. This should allow to export value added semantic meshes without having to strip out user and system data. This was also the idea behind separating countries in their own table - today, we would map this information in the topic map, and we show the separate table only for completeness' sake.

The database structure allows us to map the complete Topic Maps model, and represent each Topic Maps construct in the relational database. The result of this work can be seen in Fig. 2.

6   System Architecture

Fig. 3 shows the technology structure, fitted for the structural parts of the web application: The database is MySQL 4.1.x, the MVC paradigma is implemented using CakePHP, the browser gives us the presentation and Ajax functionality.

We use a standard current LAMP platform, extended by the use of the Cake-PHP, which allows us to have structured and enhancable code in PHP. Basic

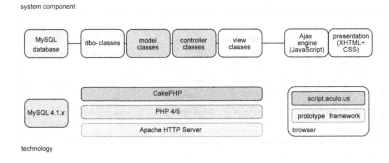

Fig. 3. Visualisation of the system architecture

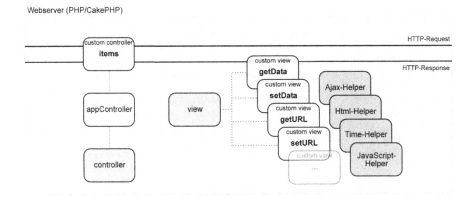

Fig. 4. An example of the interoperation of controller, view, and helper classes

classes of the Cake framework have been used to derive, amongst others, helper classes, which allow the implementation of AJAX functionality. An example of how such helper classes come to use can be seen in Fig. 4.

7 Discussion

When the decision on the underlying technologies had to be made, we focused on a lightweight and simple solution, which meant to avoid Java, and (unfortunately) TM4J [6] as a Topic Maps API with it. This meant that we had to develop our own solution for the backend, which, of course, was met with the arrival of the TMAPI for PHP [7], once we were almost finished. Today, we would naturally recommend using available solutions, which would also save lots of time (and nerves).

Next, we found that there is still room for improvement concerning data storage. We use large amounts of memory and processing ressources to be compatible to most of the XTM syntax. While this is very convenient for a prototype, as it keeps you flexible, it is also quite expensive when it comes to complex database queries and the added time resulting from the complexity. Once the extension of the prototype's functionality has settled, it is recommended to optimise the data model.

The decision for an AJAX frontend is something we are still happy with, as it allowed us to implement platform independent and sophisticated user interfaces, thus mostly minimising shortcomings of classic web applications. Those interfaces include drag and drop, context menues, and keyboard shortcuts for actions like renaming. It must be said, though, that this benefit comes with the price of essential development cost, which, especially for a project of such limited resources, has to be taken into account.

Fig. 5 shows the prototype as a user would see it in a firefox window. Currently, the user has to login onto the platform using a password, and multiuser capabilities are present in the data modeling, but sharing bookmarks with other users is not yet implemented. To the left, we see the tree, as known from other applications. In the middle, we create associations between items by dragging them from the tree onto both sides of an association type, which we then choose from a dropdown list. Finally, we determine which item plays which role, and create the association. Properties of the item can be edited on the right hand side. Plans to also include, possibly as a plugin, a graphical visualisation of parts of the underlying topic map had to be postponed, as had to be a linear history of bookmap items already visited.

An informal user evaluation has shown that users quickly grasp the elements and functionality of the user interface, and quickly achieve usual interaction goals (Search, Sort, Move, Follow Assoc., edit properties) easily. The system has been tested to import up to over 6000 items, which it manages to complete in under 30 minutes (VMWare virtual Linux server, running on Dell Latitude D610, 1,73GHz Pentium Monocore Processor), and then allows the user to interact in a fluent manner.

Fig. 5. A screenshot of the final application

While this shows that technically, the system is at least usable, a more detailed user evaluation will have to show if the users will invest the time necessary to build complex topic maps. It will have to be seen if the provided interfaces, but especially if the provided associations are capable and a big enough incentive for users to arrange, sort and organize their collection of bookmarks, which, once shared with other users, might enlighten and guide novices in a certain domain.

8   Conclusion and Outlook

We have presented the bookmap system, a web application to import, organise and present website bookmarks in both, a hierarchical and semantically associated manner. The system is a prototype and was implemented using an open source LAMP environment, extensions implemented using the CakePHP framework, PHP- and javascript. The resulting user interface is available via the Firefox web browser and delivers a rich user experience including drag and drop functionality via AJAX.

We have published the current source code on sourceforge.net [8]. Now, and depending on the feedback and input of the open source community as well as the Topic Maps community, the following topics can be next steps for bookmap: Implementation of the already prepared multi-user features; evaluation of TMAPI-PHP, and possible redesign using TMAPI-PHP; Specification, design and implementation of a visualisation plugin; Specification, design and implementation of a Firefox extension/addon to enhance and improve the desktop integration capabilities.

References

1. Passin, T.B.: Browser bookmark management with Topic Maps. In: Extreme Markup Languages (2003), (Last accessed, like all other online references, on 22. 06. 2007), `http://www.mulberrytech.com/Extreme/Proceedings/html/2003/Passin01/EML2003Passin01.htm`
2. Johannesen, A., Passin, T.B.: TM4Jscript – Topic Maps For JavaScript. Techn. Rep., Open Source Technology Group - Sourceforge.net. (2004), `http://sourceforge.net/projects/tm4jscript`
3. Biddulph, M.: Introducing del.icio.us. Techn. Rep., O'Reilly Media, Inc. - XML.com. (2004), `http://www.xml.com/pub/a/2004/11/10/delicious.html`
4. Woodworth, G.J., Masters, L.E., Putkonen, M.: CakePHP - The Rapid Development Framework. Techn. Rep., Cake Software Foundation, Inc., Nevada (2005), `http://cakephp.org/`
5. Mugnaini, L.: Mapping Topic Maps on Relational Databases. Techn. Rep. (2003), `http://www.geocities.com/xtopicmaps/mapping_xtm_on_databases.html`
6. Fröhlich, C., Ahmed, K.: TM4J – Topic Maps For Java. Techn. Rep., The TM4J Project Group (2004), `http://tm4j.org/`
7. Holje, E., Schmidt, J.: TMAPI/PHP. Techn. Rep., Open Source Technology Group (2005), `http://sourceforge.net/projects/phptmapi/`
8. Pradella, M., Hofmann, T.: bookmap - A Topic Maps-driven in-browser bookmark management system (2007), `http://sourceforge.net/projects/bookmap/`

A Theory of Scope

Lars Marius Garshol

Bouvet ASA, Oslo, Norway
larsga@bouvet.no
http://www.ontopia.net

Abstract. This paper describes an interpretation of scope based on the past few years of research on the subject. Based on this it defines two mathematical operators for scope filtering which correspond to common use cases for scope.

1 Introduction

Scope is a feature of Topic Maps used to represent the qualification of a statement. That is, the scope represents the context of validity for the statement. Any statement in Topic Maps is qualified using a set of topics, possibly empty, and this is called the scope of the statement. The empty scope is known as *the unconstrained scope*, and statements in this scope are considered to have unlimited validity within the context of a particular topic map.

Scope has been present in Topic Maps from the publication of the initial Topic Maps standard [ISO13250], but there has never been any very definite specification of its interpretation. This section presents some of the key problems that have been identified, and describes the current state of the research into these problems.

1.1 The Problem of Context

The first, and most serious, problem with scope was that its interpretation was widely felt to be underspecified. For example, the original ISO standard defined scope as follows:

> The extent of the validity of a topic characteristic assignment [...]: the context in which a name or an occurrence is assigned to a given topic, and the context in which topics are related through associations.

The issue is what "extent of validity" and "context" really means. XTM 1.0 [XTM1.0] uses essentially the same definition, and so does not provide much additional information. The problem is perhaps best approached by examining the ways in which scope is used. Below follows a list of different uses of scope gathered from specifications, papers, tutorials, and other writings on the subject over the years:

L. Maicher and L.M. Garshol (Eds.): TMRA 2007, LNAI 4999, pp. 74–85, 2008.
© Springer-Verlag Berlin Heidelberg 2008

- A name or occurrence might be valid in a particular language (multilinguality).
- A statement may derive from a particular source (provenance).
- A statement may be true according to a particular authority (opinion).
- An occurrence might be intended for a particular audience (say, "technician").
- A statement may only be true in a certain time period.
- A statement may have been inferred, instead of being actually present in the source data.

In all of these cases statements are seen as not being universally valid, but rather valid only in some context. There is also an expectation that statements in different contexts may contradict one another. Scope thus provides the ability to produce different, and potentially conflicting, views of the same subject area, with a scope being a definition of in what views a statement should be included.

Thus, the scope of a statement specifies in what contexts the statement is to be included, but the scope is not itself a context. Instead, a context is equivalent to one particular view of the topic map. It is also clear that each context is a subset of the full topic map. The question that remains is by what means the desired view of the topic map is specified. This takes us to the next problem in the interpretation of scope.

1.2   The Multi-topic Problem

Since a scope is a set of topics, it is not a given how to interpret a scope containing more than one topic. Are statements in this scope valid when all of the topics apply, or is it enough for one of them to apply? That is, if a statement appears in the scope $\{a, b\}$, does this mean that it is true only when both a and b apply, or also when only a or b apply?

The original Topic Maps standard made it clear (in clause 3.16) that the intended answer was "or". This was later viewed by the editors as a mistake. The next version of the standard, XTM 1.0, reversed the decision to make the answer "and". (XTM 1.0 is inconsistent on this point, and also stated that the interpretation was undefined.)

After the publication of these two standards there was a long phase where the model behind the two syntaxes was defined, and as part of this process substantial work was done on the interpretation of scope, primarily by Marc de Graauw. This work focused on the consequences of the "and"/"or" choice, and a number of important conclusions emerged from this work.

More and less restrictive scopes. The first observation was that if scope is interpreted as being an "or" expression then the number of contexts in which a statement is valid increases as more topics are added to its scope. That is, a statement in scope a is valid in fewer contexts than a statement in scope $\{a, b\}$, given that the latter would apply whether a or b were in the context, whereas the former would apply only in the first case.

Obviously, the opposite applies when the interpretation is "and". In this case if a statement has the scope a and b it is not enough for only a to apply; b also has to apply. So as topics are added to the scope of the statement, the number of contexts in which it applies decreases.

The unconstrained scope. Both the original Topic Maps standard and the XTM 1.0 specification had a concept of "the unconstrained scope", meaning the scope which claims that a statement is valid in all contexts. The original Topic Maps standard defined this as "the scope comprised of all topics in the topic map". XTM 1.0, however, was silent on the representation of this scope.

However, the arguments in the previous section make it clear that using the "and" interpretation of scope the least restrictive scope is the one which contains no topics at all. In other words, the representation of the unconstrained scope must be the empty set.

Similarly, in the "or" interpretation of scope, the least restrictive scope must be the biggest possible set of topics. This is clearly what ISO 13250:2000 intended by its definition, although other maximal sets of topics are imaginable.

Duplication of statements. A related question is when a statement can be considered to imply another. Imagine the same statement appearing in scopes a and $\{a, b\}$. In this case, can one of the statements be considered to imply the other?

It turns out that if the scope interpretation is "or" then clearly a is implied by a or b. In fact, in the "or" interpretation any scope containing n topics is equivalent to n statements each with one of the n topics as scope. In other words, in the "or" interpretation of scope the set representation of scope is just a shorthand for repeating the statement in different single-topic scopes.

In the "and" interpretation the situation is clearly different, in that a and b is obviously implied by a. In fact, in the "and" interpretation there is no shorthand for saying "this is valid in situations X and Y"; the only way to do it is to have two separate statements for each case.

So which is it? The TMDM [ISO13250-2] deliberately chose the "and" interpretation as being the most useful interpretation, and also the one that caused the fewest internal problems for the interpretation and representation of the standard. In the "or" interpretation it is not really clear why scope has to be a set, nor is it clear how to represent the unconstrained scope in implementations.

At the same time, Piotr Kaminsky [Kaminsky02] independently made the observation that XTM 1.0 (and the TMDM model based on it) have a built-in assumption that the correct interpretation of scope is "and". In XTM 1.0 and TMDM variant names are alternative forms of a topic name which can be used instead of the topic name in some specific context. In other words, variant names have a more restricted validity than topic names do. This is achieved by letting them inherit the scope of the parent topic name, and requiring that they add at least one topic to this scope. But this assumes that adding topics to a scope makes it more restricted, which again assumes that the interpretation of scope is "and".

There is in other words every reason to assume that "and" was the correct choice.

Structured scope. A third alternative to making a simple choice between "and" or "or" interpretations of a set of topics was to make the representation of scope more complex than just a simple set. This solution has been known as "structured scope" in the community, and several proposals have been put forward [deGraauw02] [Ahmed03].

The most complete proposal was developed at an ISO meeting in Montral in 2003 by Kal Ahmed, Ann Wrightson, and Dmitry Bogachev. It essentially turned scope into a set of nested sets, where each set was annotated with an "and" or "or" marker to define the interpretation. An extension to the XTM syntax was also proposed, as follows:

```
<occurrence>
  <scope compositor="AND">
    <topicRef xlink:href="english"/>
    <scope compositor="OR">
      <topicRef xlink:href="beginner"/>
      <topicRef xlink:href="intermediate"/>
    </scope>
  </scope>
  <resourceRef xlink:href="...."/>
</occurrence>
```

There are many equivalent ways to express the same scope in this representation, but there are well-known efficient algorithms for normalizing such expressions (that is, rewriting them to a canonical form). Thus duplicate suppression remains possible, even with this more complex representation of scope.

In the end, however, none of these proposals were adopted, for the following reasons:

- It required extensions to XTM 1.0, which the community in general wanted to avoid where possible.
- Structured scope is expensive to represent, for example in relational databases, and expensive to query.
- The TMDM model is already very complex, making the committee reluctant to complicate it further.
- Scope was at that time (and is still) rarely used, making it seem unlikely that there was any pressing need for this extension.

1.3   Scope and Inferencing

The relationship between scope and inferencing has not so far received much attention, but it is clear that the two are closely related. A few examples will suffice to demonstrate this. Consider the following topic map (in LTM syntax):

```
tm:type-instance(i : tm:instance, t : tm:type)
tm:supertype-subtype(s : tm:supertype, t : tm:subtype)
```

Here it is clearly safe to assume that i is an instance of s. But what about in this case?

```
tm:type-instance(i : tm:instance, t : tm:type) / a
tm:supertype-subtype(s : tm:supertype, t : tm:subtype) / a
```

It does not appear safe to assume that i is an instance of s in the unconstrained scope, but that this is valid in the scope of a seems unproblematic. And the following case?

```
tm:type-instance(i : tm:instance, t : tm:type) / a
tm:supertype-subtype(s : tm:supertype, t : tm:subtype) / b
```

Here it does not seem permissible to assume anything, except perhaps that i is an instance of s in the scope $\{a, b\}$. The reasoning is that under the "and" interpretation this is the minimal context in which both of the statements required for the inference are valid.

1.4   Scope and Reification

It has often been pointed out that anything which can be expressed using scope can also be expressed by means of reification. For example, if one says

```
[norway = "Norge" / norwegian]
```

this can also be expressed using reification:

```
[norway = "Norge" ~ no-no-name]
dc:language(no-no-name : dcc:subject, norwegian : dcc:value)
```

In both cases we are stating that the name is Norwegian, but in one case we are using scope, and in another we are using reification. It could be argued that the reification approach is more explicit, and it could also be argued that the scope approach is simpler and more light-weight.

The main difference between the two is that the Topic Maps technologies have better support for the scope approach (in constraints and querying) and less good support for the reification approach.

1.5   Scope and the Identity of Statements

It is worth noting that in the TMDM the scope property is included when statements are compared for equality. This means that the following topic map contains two statements:

```
[norway = "Norge" / norwegian
       = "Norge" / swedish]
```

This also means that a topic reifying one of these names will specifically reify either the Swedish or the Norwegian name, and not both. In other words: scope is part of the identity of a statement.

1.6   Other Related Work

This section has reviewed much work that is directly related to this article, but there is also a body less directly related work.

An early paper on scope in Topic Maps was [Pepper01], perhaps best known for introducing the notion of "axes of scope", which is still relevant, but tangential to the theory presented in this paper. It also presented some early examples of filtering operators.

Much work on context has also been performed in the Semantic Web community, where "named graphs" has been proposed as one solution for RDF. Important papers from this work are [?] and [Carroll05].

There is also a rich body of work on context outside semantic technologies, such as the theory f

2   The Theory of Scope

The theory of scope is formulated on the TMRM [ISO13250-5] representation of Topic Maps currently being defined by ISO. The representation is not entirely finalized yet, but is close to its final form, and updating the theory to conform to the final representation, if necessary at all, should be easy.

The theory consists of two operators which filter the topic map to leave only those statements which are true in a given context. In this paper they are based on the TMRM representation of Topic Maps, but they are not bound to it, however, and could be redefined based on any Topic Maps representation.

2.1   The TMRM Mapping of the TMDM

An understanding of this representation is essential in order to follow this section, and so this section gives a brief recapitulation of the mapping.

In the mapping, topics are represented by TMRM proxies of the form shown below. The set of all topic proxies is denoted $\mathcal{T}$.

```
{ <subject-identifier, ...>,
  <subject-locator, ...>,
  <item-identifier, ...> }
```

All statements take the form shown below. The set of all statement proxies is denoted $\mathcal{S}$.

```
{ <type, ...topic proxy...>,
  <scope, ...scope proxy...>,
  <roletype-1, ...topic proxy...>,
  <roletype-2, ...topic proxy...>,
  ...
  <roletype-n, ...topic proxy...> }
```

Scope proxies are of the following form:

```
{ <member, ...topic proxy...>,
  <member, ...topic proxy...>,
  ... }
```

Two operators from the TMRM are used in this paper. The first is $k \leftarrow_M v$, which returns all proxies which contain $\langle k, v \rangle$. The second is $p \rightarrow k$, which returns all values for the key k in the proxy p.

For convenience we also define the function $\sigma : \mathcal{S} \rightarrow 2^{\mathcal{T}}$, which given a statement produces its scope, as follows:

$$\sigma(s) = s \rightarrow \texttt{scope} \rightarrow \texttt{member} \tag{1}$$

2.2 Statements and Contexts

One assumption underlying the theory is that scope is not part of the statement, but a qualification of the statement. Consider the following statement (in LTM):

```
part-of(norway : part, denmark : whole) / eighteenth-century
```

The meaning of this is that if your context includes the eighteenth century, you can assume that Norway is part of Denmark. If it does not, you can not make that assumption.

So once a topic map has been filtered (for example using the two operators) to include only those statements which can be assumed to be true in a particular context, scope can effectively be ignored. All statements in the filtered topic map have passed the filter, and are therefore considered true. Whether, and how, the statements are qualified is then inconsequential.

2.3 Belief

In order to model belief we define the function $b(M, s)$ where is a topic map and s is a set of proxies representing the topics in which we believe. The function returns a subset of the topic map containing only the statements which are considered true in this context.

The b function must have the following properties:

- For all models M it must be the case that $b(M, \mathcal{T}) = M$. That is, if we believe everything, then all statements in the map must also be believed.
- Similarly, if we believe nothing, only statements in the unconstrained scope must be accepted. That is, for all maps M it must be the case that

$$b(M, \emptyset) \cap \mathcal{S} = \texttt{scope} \leftarrow_M \emptyset \tag{2}$$

Given the semantics of the scope set, it is clear that if a statement q has a and b in its scope, both a and b must be believed for q to be believed. That is, statements are believed iff their scopes contain no topics which we do not believe. This leads directly to the following formulation of b:

$$b(M, s) = \{q \in M | q \notin \mathcal{S} \vee \sigma(q) \subset s\} \tag{3}$$

It is easy to see that this formulation of b has the properties stated above.

2.4   Disbelief

In order to model disbelief we define the function $d(M, s)$, which produces the subset of M which could still potentially be true, but leaves out the statements which we do not believe. d should have the properties that:

- when there is nothing we disbelieve all statements in the topic map should be believed. That is, $d(M, \emptyset) = M$.
- Similarly, if we disbelieve everything, that should be the same as believing nothing, ie: $d(M, \mathcal{T}) = b(M, \emptyset)$.

The formulation of the function is easy, as it simply removes statements whose scopes contain a topic we disbelieve:

$$d(M, s) = \{q \in M | q \notin \mathcal{S} \vee \sigma(q) \cap s = \emptyset\} \tag{4}$$

Again it is easy to see that this formulation has the desired properties.

2.5   Semantics of the Operators

More and less restrictive scopes. The reasoning given above in section 1.2 on page 76 shows that the validity of a scope becomes narrower as more topics are added to it. The b and d operators need to respect this in order to accurately reflect the semantics of scope. In other words, given two statements $q, q' \in M$, if $\sigma(q) \subset \sigma(q')$ there cannot be any context c for which $q' \in b(M, c)$ but not $q \in b(M, c)$.

It's easy to see that $q' \in b(M, c)$ implies that $\sigma(q') \subset c$. Given that we know $\sigma(q) \subset \sigma(q')$, it follows that $\sigma(q) \subset c$, and therefore $q \in b(M, c)$. In other words, the b operator does honour the semantics.

A similar argument applies to d, in that assuming the same relationship between q and q' there cannot be any c for which $q' \in d(M, c)$ but not $q \in d(M, c)$. Here $q' \in d(M, c)$ implies that $\sigma(q') \cap c = \emptyset$. Given that q's scope is a subset of q''s, it's clear that the claim holds here, too.

Equivalence. A key question for the theory to answer is what sets of statements could be considered equivalent. For example, is

```
part-of(norway : part, denmark : whole) / eighteenth-century
```

equivalent to the following?

```
part-of(norway : part, denmark : whole) / eighteenth-century
part-of(norway : part, denmark : whole) / eighteenth-century lmg
```

In fact, it is. We know that any filtered topic map which contains the second association will also contain the first. Further, in filtered topic maps scopes are ignored (per section 2.2 on the preceding page), and so we know the second topic map makes the relationship between Norway and Denmark hold in exactly the same contexts as the first topic map.

Inferencing. We are now ready to revisit the question of inferencing from section 1.3 on page 77 in more depth, and show that there are two different ways to approach the problem, and that both give the same result.

The first approach is to use inference rules that ignore scope, and to apply them so that when a set of statements z in the topic map cause another set of statements z' to be inferred, the scope of all statements in z' is set to the union of the scope of all statements in z. This way, it is impossible to use the b and d operators in such a way that a filtered topic map contains an inferred statement (that is, a statement in z') without all assumed statements (that is, all statements in z) also being included.

The second approach is to filter the topic map first, and then to perform inferencing afterwards, using the remaining statements, again ignoring scope in the inferencing process itself. This also ensures that all assumptions necessary for an inferred statement are included in the topic map when the inferred statement is included.

3   Evaluation

It's not enough for the scope operators to be correct, they must also be useful, and the best test of this is whether or not they support the use cases outlined in section 1.1 on page 74.

3.1   Language

This use case is usually handled by representing each language as a topic, and attaching to each name and occurrence a scope showing what language the name/occurrence is valid in.

There are two different ways to solve this with the two operators:

– Believe only the desired language, and no other language. This will return a topic map with only text in the desired language. If no other topics are believed this means that statements in non-language scopes will be filtered out. This can be avoided by also believing other non-language topics as necessary to keep desired statements.
– Disbelieve all languages other than the desired language. This produces a topic map containing only text in the desired language without removing statements in non-language scopes. This may be easier if it is for some reason difficult to know what other scopes to believe.

3.2   Provenance

Provenance is a common name for the use case of tracking what information derives from what source. In Topic Maps this can be done by creating a topic for each source and then adding the source topic to the scope of all statements from each source.

There are several different ways this can be used. One is to believe a particular source (or set of sources) and thus see the topic map according to these sources. Another is to disbelieve a source (or, again, a set of sources) and thus remove suspect information from the topic map.

3.3   Opinion

In some cases there is a lack of agreement about facts in a domain. For example, the Scripts and Languages topic map classifies scripts into six different types of script (alphabet, abjad, abugida, featural, syllabary, logographical), following the terminology of Peter T. Daniels. However, other experts on writing systems, like William Bright, use another classification (alphabet, abjad, alphasyllabary, featural, syllabary, logographical) [Bright00]. Similarly, opinions are divided on which scripts were derived from which other scripts, etc.

This can all be expressed in Topic Maps using a scope for each school of opinion, and scoping disputed statements accordingly, as follows:

```
type-instance(devanagari : instance, abugida : type) / daniels
type-instance(devanagari : instance, alphasyllabary : type) / bright
type-instance(latin : instance, alphabet : type)
```

This again makes it possible to see the topic map according to a particular school of opinion by believing one school of opinion, or disbelieving one or more.

3.4   Audience

Relevant information resources, whether connected to subject topics via associations or occurrences, might be suitable only for particular audiences. Examples of audiences might be technicians, end-users, engineers, or managers.

Here, the goal is to show a member of a particular audience only relevant resources. This is easily done by either believing that audience, or disbelieving all other audiences. The use case is very similar to that of language.

3.5   Time

Some statements are only true in a given time period, such as:

```
part-of(norway : part, denmark : whole) / eighteenth-century
```

Again, believing one time period or disbelieving all other time periods will solve this.

3.6   Inferred Statements

One possible scope for statements is to mark them as being inferred, as opposed to being actually present in the source data. The operation users are most likely to want in this case is to view the topic map without the inferred statements. Assuming `inferred` is a scoping topic, this is easily handled with $d(M, \{\mathtt{inferred}\})$.

4   Conclusion

We have proposed a theory of scope, and shown that it is compatible with the current knowledge about scope in Topic Maps, and that it can handle the known use cases for scope.

More work remains before scope is fully understood, however, especially on the interaction between constraints and scope, but that is beyond the scope of this paper.

References

[deGraauw02] de Graauw, M.: Structuring scope; ISO SC34 working document N0347, (November 11, 2002),
http://www.marcdegraauw.com/files/structuring_scope.htm,
http://www.marcdegraauw.com/files/structuring_scope.htm

[Ahmed03] Pepper, S.: ISO/IEC JTC 1/SC34/WG3 Meeting Report (Montreal, August 1-4) ISO SC34 working document N0439, (May 7, 2003),
http://www.jtc1sc34.org/repository/0439.htm

[ISO13250-5] ISO/IEC CD 13250-5: Topic Maps – Reference Model, November 24, International Organization for Standardization, Geneva, Switzerland (2007),
http://www.jtc1sc34.org/repository/0939.pdf

[ISO13250-2] ISO/IEC IS, 0-2:2006: Topic Maps – Data Model, International Organization for Standardization, Geneva, Switzerland (2006),
http://www.isotopicmaps.org/sam/sam-model/

[ISO13250] ISO/IEC, I.S.: 0:2000: Information Technology - Document Description and Processing Languages - Topic Maps. International Organization for Standardization, Geneva, Switzerland (2000),
http://www.y12.doe.gov/sgml/sc34/document/0129.pdf

[XTM1.0] Pepper, S.; Moore, G.; *XML Topic Maps (XTM) 1.0*; August 06, 2001; TopicMaps.Org specification (2001) http://www.topicmaps.org/xtm/1.0/

[Kaminsky02] Kaminsky, P.: Integrating Information on the Semantic Web Using Partially Ordered Multi Hypersets; master of science thesis, University of Victoria (2002), http://www.ideanest.com/braque/Thesis-web.pdf

[Guha04] Guha, R., McCool, R., Fikes, R.: Contexts for the Semantic Web. In: McIlraith, S.A., Plexousakis, D., van Harmelen, F. (eds.) ISWC 2004. LNCS, vol. 3298. Springer, Heidelberg (2004)

[Bright00] Bright, W.: A Matter of Typology: Alphasyllabaries and Abugidas. Studies in the Linguistic Sciences 1(1) (Spring 2000), ISSN 0049-2388; University of Illinois at Urbana-Champaign, http://www.linguistics.uiuc.edu/sls/

[Pepper01] Peper, S., Grønmo, G.O.: Towards a General Theory of Scope. In: Proceedings of Extreme Markup (2001),
http://www.ontopia.net/topicmaps/materials/scope.htm

[Carroll05] Carroll, J.J., Bizer, C., Hayes, P., Stickler, P.: Named Graphs, Provenance and Trust; HP Labs technical report HPL-2004-57R1; (April 19, 2005),
http://www.hpl.hp.com/techreports/2004/HPL-2004-57R1.html

[Costanza05] Costanza, P., Hirschfeld, R.: Language Constructs for Context-oriented Programming – An Overview of ContextL. In: Proceedings of the Dynamic Languages Symposium (DLS), co-located with the Conference on Object Oriented Programming Systems Languages and Applications (OOPSLA), San Diego, California, USA, October 18, 2005, ACM DL (2005)
http://www.swa.hpi.uni-potsdam.de/publications/media/
CostanzaHirschfeld_2005_LanguageConstructsForContextOriented
ProgrammingAnOverviewOfContextL_AcmAuthorsVersion.pdf

Comparing Topic Maps Constraint Specification Languages

Giovani Rubert Librelotto[1], Renato Preigschadt de Azevedo[1],
José Carlos Ramalho[2], and Pedro Rangel Henriques[2]

[1] UNIFRA, Centro Universitário Franciscano, Santa Maria, RS, 97010-032, Brasil
{librelotto,rpa.renato}@gmail.com
[2] Universidade do Minho, Departamento de Informática
4710-057, Braga, Portugal
{jcr,prh}@di.uminho.pt

Abstract. Topic Map Constraint Language (TMCL) provides a means to express constraints on topic maps conforming to ISO/IEC 13250. In this article, we will use a test suite and show, step-by-step, the way we handled several kinds of Topic Maps constraints in many different instances in order to answer questions like: Do they do the same job? Are there some kinds of Topic Maps constraints that are easier to specify with one of them? Do you need different background to use the tools? Is it possible to use them in similar situations (the same topic maps instances)? May we use them to produce an equal result? How do AsTMa!, OSL, Toma, and XTche relate to Topic Maps Constraint Language (TMCL)? What kind of constraints each one of these three can not specify? We will conclude this paper with a summary of the comparisons accomplished between those Topic Maps constraint languages over the use case proposed.

1 Introduction

Topic maps are an ISO standard for the representation and interchange of knowledge, with an emphasis on the findability of information. A topic map can represent information using topics (representing any concept), associations (which represent the relationships between them), and occurrences (relationships between topics and information resources relevant to them). They are thus similar to semantic networks and both concept and mind maps in many respects.

According to Topic Map Data Model (TMDM) [GM05], Topic Maps are abstract structures that can encode knowledge and connect this encoded knowledge to relevant information resources. On one hand, this makes Topic Maps a convenient model for knowledge representation; but on the other hand, this can also put in risk the topic map consistency. A set of semantic constraints must be imposed to the topic map in order to grant its consistency.

Currently, we can find three approaches to constrain Topic Maps – OSL [Gar04], AsTMa! [Bar03], Toma [Pin07], and XTche [LRH] – that allow us to specify constraints and to validate the instances of a family of topic maps against that set

L. Maicher and L.M. Garshol (Eds.): TMRA 2007, LNAI 4999, pp. 86–97, 2008.

of rules. With these resemblances it is easy to conclude that they are quite similar. However they differ in some fundamental concepts. These three Topic Maps constraint specification languages were thoroughly tested and benchmarked with a huge test suite. The most significant results will be discussed in this paper.

The paper is organized as follows. In the next section, the Topic Maps Constraint Languages are introduced. The used case study – an e-Commerce corporation – is introduced in section three. Section four presents the comparison among the main constraint languages. Finally, conclusions are given in section five.

2   Topic Maps Constraint Language (TMCL)

Given a specification, a constraint is a logical expression that restricts the possible values that a variable in that specification can take.

A domain specific language enables the description of constraints required by each problem in a direct, clear and simple way; moreover it allows the derivation of a program to automatize the validation task. The derived semantic validator will verify every document, keeping silent when the constraints are satisfied, and reporting errors properly whenever the contextual conditions are broken.

The language to define topic map constraints is called as Topic Map Constraint Language. This language is currently on its way for standardization (ISO 19756 [NMB04]). The objective of TMCL is to allow formal specification of rules for topic map documents. TMCL has a similar purpose as schema languages for relational databases or XML applications. The constraint language is required to formalize application of specific rules. Currently there are different proposed constraint languages that will be presented in the next subsections.

2.1   XTche Language

XTche is a process for specifying constraints on topic maps with a constraint language. This language allows to express contextual conditions on classes of Topic Maps. With XTche, a topic map designer defines a set of restrictions that enables to verify if a particular topic map is semantically valid.

XTche is an XML Schema oriented language [DGM+01]. This fact brings two benefits: on one hand it allows for the syntactic specification of Topic Maps (not only the constraints); and on the other hand it enables the use of an XML Schema editor (for instance, XMLSpy[1]) to provide a graphical interface and the basic syntactic checker.

The constraining process is composed of a language and a processor [LRH]. The language is based on XML Schema syntax. The processor is developed in XSLT language. The XTche processor takes a XTche specification and it generates a particular XSLT stylesheet. This stylesheet can validate a specific topic map (or a set of them) according to the constraints in the XTche specification.

---

[1] http://www.altova.com

XTche language meets all the TMCL requirements [NMB04]; for that purpose, XTche has a set of constructors to describe constraints in Topic Maps. But the novelty of the proposal is that the language also permits the definition of the topic map structure in an XML Schema style. An XTche specification merges the schema (defining the structure and the basic semantics) with constraints (describing the contextual semantics) for all the topic maps in that family.

2.2   AsTMa! Language

AsTMa! [Bar03] is another language for constraint topic maps aiming to validate topic maps against a given set of rules. AsTMa! is a member of AsTMa* family (which includes AsTMa= for authoring TM, and AsTMa? for querying TM) and exposes some features of TMCL, because it has written earlier than the final version of the TMCL.

Resembling XTche the constraining process is composed of a language and a processor. The language is based on the Perl language, and the processor is written in Perl. At this time the AsTMa! Language is no longer maintained, because the author is working on a completely new distribution. So for this article we assume the AsTMa! Language definition for expression evaluation.

2.3   OSL Language

According to the Ontopia Schema Language specification [Gar04], OSL has been designed to have a minimal number of features available on TMCL and a minimum expressive power.

Basically the OSL language constraints only the structure of a Topic Map. An OSL schema consists of a set of topic and association class definitions. These class definitions constrain the structure of the instances of the classes, and so control the form information may take in a topic map that uses the schema [Gar04].

As the other two languages OSL is composed by a language and a processor. The language is based on XTM [PH03] and the processor is written in Java, available for running standalone or as a plug-in for Ontopia Omnigator [Ont02].

2.4   Toma Language

Toma [Pin07] is a TMCL language with a syntax very similar to SQL. It is based on path expressions that allow to access all elements in the topic map. This language has constructors like SELECT, UPDATE, INSERT, and DELETE; this constructors can be used to query and manipulate the topic map.

This language offers the MERGE statement to enable topic maps fusion. Also it offers the EXPORT statement to export the topic map to XTM. Toma provides functions that allow to modify, convert and aggregate data coming from the topic map. So, this language can be used as a Topic Maps Query Language (TMQL) and Topic Maps Manipulation Language (TMML).

3 Case Study – E-Sell Corporation

A list of requirements for the new language was established by the ISO Working Group – the ISO JTC1 SC34 Project for a TMCL [NMB04]. Part of this document, consists in a section that presents a case study for a language to constraint any topic map. This case study is about an e-Commerce application.

For this use case, we created a topic map that stores information about customers, products and orders made by customers.

From that we will formalize the design decisions we have taken, and specify its vocabulary and type system (taxonomy). After that we will add application specific rules in AsTMa!, XTche, Toma, and OSL, the constraining languages.

The E-Sell's Ontology: The objective of the ontology is to define sets of vocabularies along with its meaning that will be used within the framework. Rules or constraints also need to be defined to ensure the rigidity of the Topic Map framework that is used. This ensures that the information contained within the document is valid.

From the product class we derive subclasses which are the categories of products like: beverage, technology, and clothing. Some topics of type "product" like: wine, radio, television, DVD, and phone are created. Another topic that needs to be covered as a class is the customer. From this class we derive subclasses which are the different customer categories like person and company.

Figure 1 shows a graph that represents a small part of E-Sell's ontology on Vizigator [Gen]. This figure presents main topic types (order, product, and customer), the other topic types (person, company, technology, and beverage), and the topic instances (order 01, radio, Ronnie Alves, ...).

The links in that figure show the relationship between topic type and topic instances (beverage and wine, for instance) or association between two (or more) topics (for instance, order 05 is composed of DVD, radio, television, and wine).

4 Comparing the Topic Maps Constraint Languages

In this section we will compare the three languages briefly described previously in this paper. Then we point out advantages and disavantages of eachone.

4.1 Do You Need Different Background to Use the Languages?

Yes. To use the XTche language, the topic map designer needs to have solid understanding about XML, XML Schema, XSLT, and XPath [CD99]. All XTche specifications are in XML Schema format, so the designer can use a visual tool to write the constraints. The constraint can be written in any text editor, but it has the complexity of a common XML Schema.

To specify AsTMa! constraints, the designer is required to know the AsTMa! particular syntactic. To run the AsTMa! processor, it is necessary to have Perl and Prolog compilers installed.

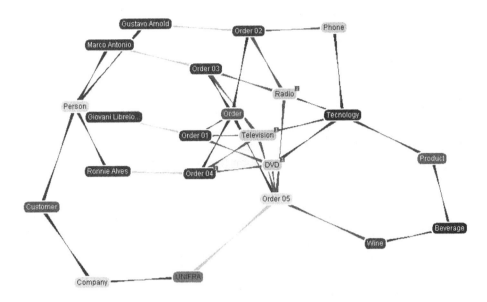

Fig. 1. The E-Sell Corporation's ontology

Toma language is SQL-based. It also takes some ideas from Object Orientation notation, Tolog language, and AsTMa* syntax.

OSL language is XTM-based, so the designer needs to specify this kind of constraints in agreement with XTM elements. The OSL tool only requires support for the Java language. Another way to execute OSL verifications is running it on Omnigator [Ont02].

4.2   Do They Do the Same Job?

Not really. To illustrate this subject, we will present a few comparisons among these languages.

Validating generic topic map structure: In the first example, XTche, OSL, Toma, and AsTMa! languages virtually do the same job. These three languages allow to verify if a topic map (or a family of topic maps) has some inconsistency in agreement with a set of rules about its structure and content.

For instance, the association *is-making-order* represents each product line. This creates a relationship between a particular *product* and an *order*, along with the *quantity* of the product ordered. It means that an association of type *is-making-order* must have three association roles: *product*, *order*, and *quantity*. The code below shows the AsTMa! specification:

```
forall $a [ (is-making-order) ] => exists $a ] (is-making-order)
        product: *
        quantity: *
        order: * [
```

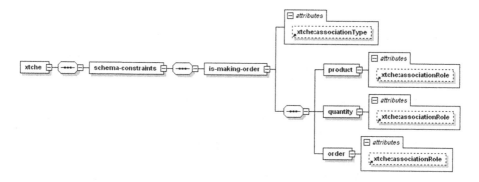

Fig. 2. XTche specification for an association structure

Figure 2 shows a graphical representation of this constraint specified in XTche. According to the Toma language, the same cosntraint is specified as shown below:

```
define constraint is_making_order_constraint
 each association $a(is-making-order)
 satisfies exists $a(is-making-order)->product = $$
   and $a(is-making-order)->quantity = $$
   and $a(is-making-order)->order = $$;
```

In OSL language, the same specification is:

```
<association>
    <instanceOf>  <internalTopicRef href="#is-making-order"/>  </instanceOf>
    <role min="1" max="1">
      <instanceOf>  <internalTopicRef href="#product"/>  </instanceOf>
      <player>  <internalTopicRef href="#product"/>  </player>
    </role>
    <role min="1" max="1">
      <instanceOf>  <internalTopicRef href="#order"/>  </instanceOf>
      <player>  <internalTopicRef href="#order"/>  </player>
    </role>
    <role min="1" max="1">
      <instanceOf>  <internalTopicRef href="#quantity"/>  </instanceOf>
      <player>  <any/>  </player>
    </role>
</association>
```

Validating a specific topic map structure: In the second example, the constraint is also about the association *is-making-order* where we need to ensure that a topic instance of *product* plays the role *product* and a topic instance of *order* plays the role *order*.

The code below introduces the AsTMa! function called *exists*:

```
forall $a [ (is-making-order) ]
=> exists $a ] (is-making-order)
        product: $p
        quantity: *
        order: $o [
    and
    exists [ $p (product) ]
    and
    exists [ $o (order) ]
```

In XTche, the same specification would be:

```
<xs:element name="schema-constraints">
  <xs:complexType>  <xs:sequence>
      <xs:element name="is-making-order">
        <xs:complexType>  <xs:sequence>
            <xs:element name="product">
              <xs:complexType>
                <xs:sequence>
                  <xs:element name="product">
                    <xs:complexType>
                      <xs:attribute ref="xtche:associationPlayer"/>
                      <xs:attribute ref="xtche:topicType"/>
                    </xs:complexType>
                  </xs:element>
                </xs:sequence>  <xs:attribute ref="xtche:associationRole"/>
              </xs:complexType>
            </xs:element>
            <xs:element name="quantity">
              <xs:complexType>
                <xs:attribute ref="xtche:associationRole"/>
              </xs:complexType>
            </xs:element>
            <xs:element name="order">
              <xs:complexType>  <xs:sequence>
                  <xs:element name="order">
                    <xs:complexType>
                      <xs:attribute ref="xtche:associationPlayer"/>
                      <xs:attribute ref="xtche:topicType"/>
                    </xs:complexType>
                  </xs:element>
                </xs:sequence>  <xs:attribute ref="xtche:associationRole"/>
              </xs:complexType>
            </xs:element>
          </xs:sequence>  <xs:attribute ref="xtche:associationType"/>
        </xs:complexType>
      </xs:element>
    </xs:sequence>
  </xs:complexType>
</xs:element>
```

The diagrammatic view of this schema (and the next ones too) can be generated by any XML Schema editor, so we will not show them in the paper.

Toma language defines this constraint like the following:

```
define constraint is_making_order_constraint_testing_product
 each association $a(is-making-order)
 satisfies exists $a(is-making-order)->product = $p
   and $a(is-making-order)->quantity = $$
   and $a(is-making-order)->order = $o
   and $p->type = 'product'
   and $o->type = 'order';
```

Unfortunately, OSL language does not allow to specify this kind of constraint.

Data types: According to the TMDM [GM05], Topic Maps do have a concept of data and data types, but there is no commitment to any set of primitives such as XML Schema (XSD) [DGM+01]. That may be a good move, since XSD is – like any other set – quite arbitrary. Useful, but arbitrary. So, if a topic map designer wants to validate an age occurrence as a number, he needs to use a constraint language.

The only way to constrain text in AsTMa! is to use regular expressions. For instance, to allow the invocation of "boolean test functions", such as:

```
in (age): ?is_age()
```

The AsTMa! validator would call this function (implemented externally). According to its creator, this issue would have to be addressed if AsTMa! evolves into a new version.

XTche specification below tests if a person type topic has an age occurrence of integer type (any XSD data type can be used in a XTche specification).

```
<xs:element name="schema-constraints">
  <xs:complexType> <xs:sequence>
      <xs:element name="person">
        <xs:complexType>
          <xs:sequence>
            <xs:element name="age">
              <xs:complexType>
                <xs:simpleContent>
                  <xs:extension base="xs:integer">
                    <xs:attribute ref="xtche:occurrenceType"/>
                  </xs:extension>
                </xs:simpleContent>
              </xs:complexType>
            </xs:element>
          </xs:sequence> <xs:attribute ref="xtche:topicType"/>
        </xs:complexType>
      </xs:element>
    </xs:sequence>
  </xs:complexType>
</xs:element>
```

Toma has functions to convert parameters. The function *to_num* converts text to a number. The function *to_unit* converts between units defined by *Units Conversion Library* by *Maio Fundation*[2]. However, this language does not allow the user to use all XSD data types as XTche does.

In terms of data type, OSL does not provide any kind of data type.

4.3   Where Each Language Shows Its Strength

The topic maps constraints about topics and association structures are easier to specify in these four languages. For instance, the constraint "customer must have a contact number which is either a phone or a fax number" is specified in AsTMa! like this:

```
forall $c [ * (customer) ]
=> exists $c [ in (phone): * ]
   or
   exists $c [ in (fax): * ]
```

According to the XTche language, the respective specification is below.

---

[2] http://sourceforge.net/docman/display_doc.php?docid=2810&group_id=19449 #SECTd0e169

```
<xs:element name="schema-constraints">
  <xs:complexType> <xs:sequence>
    <xs:element name="customer">
      <xs:complexType>
        <xs:choice>
          <xs:element name="phone">
            <xs:complexType>
              <xs:attribute ref="xtche:occurrenceType"/>
            </xs:complexType>
          </xs:element>
          <xs:element name="fax">
            <xs:complexType>
              <xs:attribute ref="xtche:occurrenceType"/>
            </xs:complexType>
          </xs:element>
        </xs:choice> <xs:attribute ref="xtche:topicType"/>
      </xs:complexType>
    </xs:element>
  </xs:sequence>
  </xs:complexType>
</xs:element>
```

The code below shows the Toma specification for this kind of constraint:

```
define constraint customer_must_have_a_contact_number_constraint
 each topic $t
   where $t.type = 'customer'
 satisfies exists $t.oc.id = 'phone'
   or $t.oc.id = 'fax';
```

In other hand, OSL correspondent code is presented below. However, this language have a limitation: it does not work with boolean operations. So the constraint "either a phone or a fax number" is not supported.

```
<topic>
    <instanceOf>
      <internalTopicRef href="#customer"/>
    </instanceOf>
    <occurrence min="0" max="1">
      <instanceOf>
        <internalTopicRef href="#phone"/>
      </instanceOf>
    </occurrence>
    <occurrence min="0" max="1">
      <instanceOf>
        <internalTopicRef href="#fax"/>
      </instanceOf>
    </occurrence>
</topic>
```

The code above defines a topic instance of customer that has zero or one phone occurrence and zero or one fax occurrence. But, according to this OSL specification, there is no way to verify if a topic instance of customer has both occurrences.

4.4   Is It Possible to Use Them in Similar Situations (The Same Topic Maps Instances)?

It is possible to use them in several similar situations but it is important to care about the topic map format. XTche language only processes topic maps in

XTM format. There is a small project in the XTche context to create a processor that converts other topic maps formats – LTM (Linear Topic Map) [Gar02] and HyTM (HyTime Topic Maps) [NBB03] – to XTM.

In the same perspective, AsTMa! language only processes topic maps that are in AsTMa= format.

Toma can not generate TM or XML content. Toma assumes that users that want to create applications using Toma will use in addition other technologies (Java, Perl, Python, etc.). Each of those technologies provide sets of techniques and methodologies to create XML content as well as any other content.

Talking about OSL, this language is part of Omnigator tool [Ont02]. Many Topic Maps formats can be validated according to a set of OSL rules. Ontopia enables the navigation over the following topic map formats: XTM, LTM, and HyTM; ontologies in RDF (Resource Description Framework) [LS99] format can be navigated by Omnigator too.

4.5   May We Use Them to Produce an Equal Result?

Maybe. The answer is *Yes* if the topic map designer wants to validate the topic map schema because these three languages confirm the validity of a topic map instance across a set of rules. The answer is *No* if the topic maps designer wants to validate the topic map with particular constraints, like existence, boolean, and conditional constraints. In this case, XTche, Toma, and AsTMa! have constructors to specify that; OSL has not.

For example, the constraint "for all topic that has the topic type customer, it must have a basename (for customer name), an occurrence (for address), a subject identifier (for customer id), and optionally additional occurrence (for email address)" [NMB04] can be constrained in XTche, AsTMa!, Toma, and OSL. The result for all these languages is a list of the topics that are not conformed with this rule. If all the topics conforms this rule, the result is the topic map validation confirmed. So, for this case: *Yes*, we may use them to produce an equal result.

However another constraint example: "for all association of is-making-order type, it must have the association roles customer and order played by the topic that is of type customer and order respectively" [NMB04] can be validated by Toma, XTche, and AsTMa! languages, and can not be validated by OSL language. Thus, for this case: *No*, we may not use them to produce an equal result.

4.6   How Do AsTMa!, OSL, Toma, and XTche Relate to Topic Maps Constraint Language (TMCL)?

Toma, XTche, and AsTMa! languages are based on a draft version of the TMCL, so they are able to specify almost any kind of constraint suggested by TMCL requirement. Toma and AsTMa! do not have constructors to constrain data types.

Toma, AsTMA! and XTche have constructors to make complex conditional, boolean, and existential constraints. On the other hand, OSL does not have relationship with TMCL, and it was defined to make just simple validations in a topic map. So the language does not have boolean, existential, and conditional operators, becoming a real alternative only in simple and small projects.

OSL was not designed on the basis of TMCL requirements; it is intended only for validating the topic maps structures. For instance, OSL can not specify the following constraint: "topic radio can not be used to scope association".

5 Conclusion

This paper showed a comparison among the three TMCL-based languages – AsTMa!, OSL, Toma, and XTche – over several kinds of Topic Maps constraints in many different instances. We started with our strong motivation to check a topic map for syntactic and semantic correctness - as a notation to describe an ontology that supports a sophisticated computer system (like the applications in the area of Semantic Web or archiving) its validation is crucial!

In order to compare these languages, we succeeded in applying a case study – E-Commerce Application (subsection 6.1 of TMCL Requirements [NMB04]) – virtually representative of all possible cases. This means that: on one hand, we were able to describe the constraints required by each problem in a direct, clear and simple way; on the other hand, the Topic Maps semantic validator could process every document successfully, that is, keeping silent when the constraints are satisfied, and detecting/reporting errors whenever the conditions are broken.

Doing a comparison among these languages, some advantages of XTche emerge: (1) XTche has a XML Schema-based language, a well-known format; (2) XTche allows the use of an XML Schema graphic editor, like XMLSpy. In a diagrammatic view, it is easy to check visually the correctness of the specification; (3) XTche gathers in one specification both the structure and the semantic descriptions, and it realizes a fully declarative approach requiring no procedural knowledge for users.

The main problem about XTche is the size of this code. If a topic map designer does not have a XML Schema editor, the specification is too complex in a comparison with other languages. This XTche problem is a Toma and AsTMa! advantage: the size of AsTMa! constraints are small, very similar to regular expressions and SQL, respectively.

In a related work, Eric Freese [Fre02] says that it should be possible to use the DAML+OIL language to provide a constraint and validation mechanism for topic map information. The cited paper discusses how to describe validation and consistency of the information contained in Topic Maps using DAML+OIL and RDF, showing how to extend XTM and how to define PSIs and class hierarchies, as well as to assign properties to topics.

The main conclusion is that XTche, Toma, and AsTMa! comply with almost all requirements stated for TMCL whereas OSL just includes topic maps structure validation.

References

[Bar03]   R. Barta.: AsTMa! Bond University, TR., (2003)
           http://astma.it.bond.edu.au/constraining.xsp

[CD99]   Clark, J., DeRose, S.: XML Path Language (XPath) - Version 1.0.
           (November 1999), http://www.w3.org/TR/xpath

[DGM+01]   Duckett, J., Griffin, O., Mohr, S., Norton, F., Ozu, N., Stokes-Rees, I.,
           Tennison, J., Williams, K., Cagle, K.: Professional XML Schemas. Wrox
           Press (2001)

[Fre02]   Freese, E.: Using DAML+OIL as a Constraint Language for Topic Maps.
           In: XML Conference and Exposition 2002. IDEAlliance (2002),
           http://www.idealliance.org/papers/xml02/dx_xml02/papers/
           05-03-03/05-03-03.html

[Gar02]   Garshol, L.M.: LTM – The Linear Topic Map Notation. Ontopia (2002),
           http://www.ontopia.net/topicmaps/ltm.html

[Gar04]   Garshol, L.M.: The Ontopia Schema Language – Reference Specification
           (2004), http://www.ontopia.net/omnigator/docs/schema/spec.html

[Gen]   Gennusa, P.: Ontopia's Vizigator(tm) - Now you see it!. In: XML 2004
           Conference and Exposition. IDEAlliance, Washington D.C., U.S.A (2004)

[GM05]   Garshol, L.M., Moore, G.: Topic Maps – Data Model. ISO/IEC JTC
           1/SC34 (January 2005),
           http://www.isotopicmaps.org/sam/sam-model/

[LRH]   Librelotto, G.R., Ramalho, J.C., Henriques, P.R.: Constraining topic
           maps: A TMCL declarative implementation. In: Extreme Markup Lan-
           guages 2005: Proceedings, IDEAlliance (2005)

[LS99]   Lassila, O., Swick, R.R.: Resource Description Framework (RDF) Model
           and Syntax Specification. In: World Wide Web Consortium (February
           1999), http://www.w3.org/TR/REC-rdf-syntax

[NBB03]   Newcomb, S.R., Biezunski, M., Bryan, M.: The HyTime Topic Maps
           (HyTM) Syntax 1.0. ISO/IEC JTC 1/SC34 N0391 (2003),
           http://www.jtc1sc34.org/repository/0391.htm

[NMB04]   Nishikawa, M., Moore, G., Bogachev, D.: Topic Map Constraint Language
           (TMCL) Requirements and Use Cases. ISO/IEC JTC 1/SC34 N0548
           (2004), http://www.jtc1sc34.org/repository/0548.htm

[Ont02]   Ontopia. The Ontopia Omnigator (2002)
           http://www.ontopia.net/omnigator/

[PH03]   Park, J., Hunting, S.: XML Topic Maps: Creating and Using Topic Maps
           for the Web. Addison-Wesley, Reading (2003)

[Pin07]   Pinchuk, R.: TopiWriter User Manual - Toma (2007)

Knowledge-Oriented Middleware Using Topic Maps

Robert Barta

rho information systems
rho@devc.at

Abstract. In this work we present an architectural overview over a knowledge-based middleware based on Topic Maps. In that, we exploit the TM paradigm as much as possible trying to solve typical middleware-specific tasks. Hereby we use the concept of virtualization to homogenize the data landscape, the TM query language for semantic transformations and a dedicated, RESTful protocol for TM fragment interchange.

1 Introduction

Middleware for distributed systems has been traditionally created along paradigms governing the programming model for these applications, be that purely RPC-based (DCE, DCOM, RMI), message-based, object-based (e.g. CORBA) or object-oriented (J2EE). If the main paradigm inside the application itself shifts more towards treating information organized into knowledge networks, any middleware has to adopt this view to be of any use. This shift itself is not novel and has been proposed for a number of application scenarios (e.g. [12]), many of those revolving around the Semantic Web [7].

Our setting here is specific in that we concentrate on the syndication of assertional knowledge, something which we shortly touch on in section 2. This presentation focuses on architectural considerations how various Topic Maps (TM, [1]) technologies can be combined and reinterpreted to form a robust and flexible middleware for these knowledge-centric applications. For this purpose we propose a layered architecture as in Fig. 1.

The task of the virtualization layer (section 4) is to abstract away from the particularities of the resource and to provide a *topic-mappish* view to the underlying data. To motivate that, section 3 discusses first a content framework for this process.

In principle, applications could then directly access any information via that layer. Instead we choose to introduce another abstraction layer for knowledge engineering purposes. The task of the *knowledge layer* (section 5) is to be able to reorganize the topic map content: Maps can be combined (merged), transformed and filtered. Additionally a service-oriented interface allows the communication with peer knowledge layers (section 6). At the end we will summarize some architectural observations and point out future research directions.

L. Maicher and L.M. Garshol (Eds.): TMRA 2007, LNAI 4999, pp. 98–115, 2008.

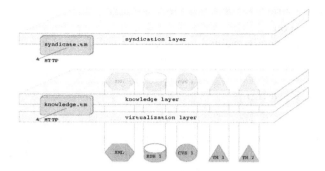

Fig. 1. Syndication Architecture

2   Syndication

While syndication is mostly associated with traditional news-carrying protocol formats, such as RSS and Atom, we use the term in the sense of the original business model. Accordingly, a syndicator is positioned between one or many producers and one or many consumers. On the *upstream side* the producers deliver content in various formats, timing and semantic quality into the syndicator. From a flow perspective, the syndicate pulls content or content is pushed into the syndicate.

The value-adding aspect of the syndicator lies in the quality assurance of the incoming content and in the repackaging of content from various sources. These content packages are then delivered in defined quality, quantity and timing to downstream consumers. Again, several technologies are at the disposal here, depending on whether the partners are individuals, web portals or another syndicator.

A quite important business case for the syndicator is to keep all or part of the traversing information in an archive. That can be used later to enrich outgoing information or to be more effective in routing specific information to particular customers.

Most syndicators nowadays operate mostly document-centric, hosting images, publications or articles. With the introduction of semi-structured documents (such as XML) a more fine-grained and targeted processing can be deployed. To guide this even further, meta data can be attached to documents and fragments thereof. Gradually, this meta data can be expanded into a semantic network, so that the document-centric view gives way to a network-centric information architecture in which the documents are just nodes within the network.

A *semantic syndicator* then handles not (only) documents but effectively semantic networks, i.e. knowledge fragments. Still, we will use the slightly imprecise term *meta-data syndication* for this setup.

3   Content Framework

As the virtualization layer is supposed to handle a wide variety of data resources to be uplifted into the TM paradigm, we first have to classify them to arrive at a content framework.

Every modern enterprise maintains its core data in some database, most commonly a relational one. In the world of database designers, *all* relevant enterprise information should be maintained in such structured form. This is then also the main objective of conventional Enterprise Information Integration (EII).

3.1 Content Paradigms

The reality, though, is that much of the information within an organisation is document centric, and not data centric. Therefore document management systems and content management systems have been the preferred solutions here. These systems may or may not use a relational store at their base; what is important, is the *structure of the data* and the *variability of that structure* within the data corpus. How the data is stored technically and via which paradigms it can be addressed, is a separate issue.

This dichotomy between document and data centricity has been partly addressed with XML [6]. With it tree-oriented structures can be described. Documents could always be naturally mapped into tree form, with chapters, sections, etc. being nodes and the text being the leaves of this tree. Tree-structures can also capture table-oriented information stored in relational databases. This has put the spotlight on transformation and query languages such as XQuery [2] and XSLT which then allow to access a wide range of information and quite flexibly transform it into other trees, tables or flat text.

Ignoring XML's idiosyncrasies, one main problem with tree-oriented structures is that they are exactly this: tree structured. As such they can perfectly host narrative information (as text in documents) and iterative content (as data in tables), but not arbitrarily structured information.

This becomes problematic when (a) within one XML document one piece of information has to reference another, or (b) one XML document has to reference information within another XML document. While this indicates that the content intrinsically has a graph structure, XML authors have to emulate this by using XML IDs or even resort to XLink to implement references. This is the main problem with *meta data*: Some formats mandate that it has to live *inside* the documents, others allow it to be hosted elsewhere.

Moreover, nothing within the XML framework itself caters for the addressing of information, or more generally, the addressing of the things the information is about. Also nothing about the semantic quality can be expressed in XML. All this has been the realm of knowledge-oriented systems all along. Many of them have *identity* built in and provide ontological frameworks, one way or the other.

Accordingly, a more generalized picture presents itself as in Fig. 2:

The first row lists the main *content paradigms*, i.e. that for relational data, object-oriented information, then trees as they are used in file systems, or XML. With increasing flexibility we then move to graphs and on the very right we see fulltext.

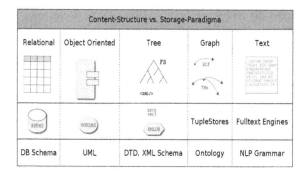

Fig. 2. Virtualization

3.2   Storage Paradigms

For each of these content paradigms dedicated technologies have been created over the last 30+ years. RDBMSes are perfectly suited for relational data, OODBMses are tailored for objects, specifically those used in OO programming languages. Nowadays dedicated XML stores are on the market, so that we can expect to see more technologies in the future specifically designed to hold graphs (often referred to as 'Tuple Stores'). The middle row in Fig. 2 also lists fulltext engines: They are specifically designed to hold fulltext documents, possibly enhanced by meta data to tag and classify documents.

That each of these technologies is created with a particular paradigm in mind has the advantage that any paradigmatic structure in the data can be hard-coded in order to achieve the best performance. A relational database system will have *table* and *row* as a first-class concept as ontological commitment. Consequently, the software can rely on the fact that every row in the table has exactly the same structure. Other structural commitments have to be made for other storage paradigms.

3.3   Crossovers vs. Native Stores

The fact that different technological commitments have to be done for each paradigm does not rule out the option to host certain information using a storage technique which has not been aligned to the same paradigm. For many years programmers use relational databases to *park* their objects. And while object-oriented databases never made it into the mainstream market, XML stores are more and more used to store object instance data, especially in the advent of SOA. There have also been massive investments in persisting XML structures within a relational database; the same holds true for storing graph information (specifically for RDF and Topic Maps) and also fulltext within these.

If the data paradigm corresponds to that of the storage technology, then this is called a *native store*. In all other configurations where there is no perfect fit, there is an impedance price to pay, be it via the added complexity of the mapping or be it with lost optimization opportunities. Mostly it is a tactical decision within an organization which storage technology is preferred. The impedance mismatch is simply traded-off with the price to maintain different databases, be that technology-wise or licence-wise.

3.4 Schemas

For each chosen data paradigm there still exist many levels of freedom. If your data is structured along connected tables, then the individual names and data-types for the columns and the nature of the connection between tables has to be detailed.

For a relational database this is done using a *database schema language*, either simply by providing CREATE SQL statements, by entity-relationship diagram formalisms. For OO databases no such clear-cut candidate languages exist, with maybe IDL and UML the most widespread in use. Also for XML a number of schema languages exist; DTD, RelaxNG, XML Schema and Schematron all have their clientel.

The situation with full-text is more complicated. The documents themselves might have a clearly defined structure; here XML schema languages can be reused. The full-text meta data (author, title, keywords, etc.) can be managed within any of the paradigms. But grammars for the natural-language text themselves are normally out of reach; here other linguistic methods must be applied to constrain and validate text.

When using graphs to capture factual content, means must exist to (a) constrain and (b) describe the form of such graphs in order to validate them against a schema. These constraints can be formalized as facts, i.e. statements about statements. This is achieved with constraint languages. The graphs, together with their constraints hold information about a particular knowledge domain. It is only consequent to enhance this basic knowledge with some degree of inferencing to allow an engine to derive further knowledge from the existing. This is achieved by extending the constraint language into an ontology definition language.

4 Virtualization

If applications are supposed to experience all data along the TM paradigm, then either that data has to be authored as topic maps in the first place, or data residing in non-TM resources have to converted at some stage. Converting large quantities of content in existing data stores into topic maps is highly unrealistic, for technical and techno-political reasons. The computational costs of such conversions are simply too high and existing database technologies exhibit a performance and functionality range which cannot be easily reproduced with a TM-only solution. Instead, TM middleware will use the mechanism of *non-materialized views*.

4.1   Virtual Maps

Accordingly, *virtual maps* only exist conceptually [3]: Whenever such maps are accessed, they will be materialized in whole or in part, depending on how the access pattern mandates it. This is the task of a *resource wrapper*.

The actual data can then be hosted in a resource which is most appropriate for the data structure: RDBMS for tabular data, XML databases for hierarchically structured documents or, say, DNS for the mapping of Internet host names to IP addresses. A resource-dependent wrapper will encapsulate the resource and provides so a "TM-ish view" (Fig. 3) to the environment.

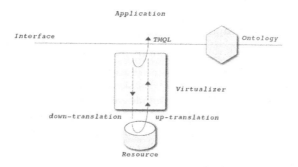

Fig. 3. Virtualization

Whenever the application now uses either a TM API or a TM query language to access the virtual map, it is then the wrapper's task to mediate between TMQL queries on the one side and the native access method on the other. For a relational resource this most likely would involve SQL. Here query results are translated by the wrapper into topics and associations before they are handed to the application.

Which structure these topic map fragments have, which taxomometric concepts are used hereby, and which types the occurrence values have, all this is defined in an ontology. If, for example, a relational database is wrapped that way, the ontology mirrors the database schema; if an XML document is wrapped, then elements in the ontology might correspond to parts in the XML schema definition for that document.

On the one hand, the ontology so becomes a formal definition of the information which can be found using the interface of the wrapper. Applications can query the ontology to learn about the interface, providing a certain degree of *reflection* in our architecture. On the other hand, the taxonometric part of the ontology and the virtual map can be considered to be merged together, so that applications have a seamless access.

4.2   Non-materializable Maps

Virtualization is not only a technique to integrate data resources with significant volume or with a high update frequency. It also works for data resources which are not materializable in the first place.

One such example is the global DNS (Domain Name Service). It is inherently distributed and there is no concept of a global consistent state as DNS records may change anytime anywhere. Still it is possible [3] to create a wrapper to provide a TM view to applications. A query for *"all IP addresses for the name www.example.org"* will return these in form of topics.

The example, however, shows also some limitations as not all access patterns which TMQL theoretically would allow are achievable. A query for *"all topics of type IP address for the DNS zone example.org"* might fail as the underlying DNS infrastructure might be unwilling to perform a zone transfer. Other wrappers may impose further restrictions on certain read or write access patterns.

4.3   Materializable Maps

Materializable maps are those which can be manifested as whole, either in memory or on disk. We regard *materializable maps* as a special case of *virtual maps*.

Still, a large quantity of maps will be materializable, notably those authored manually. For these the middleware provides a variety of *drivers*. Simple drivers allow maps to be loaded from and stored to BerkeleyDB files on the application's request. More sophisticated ones just tie the in-memory representation to the file content without actually loading the whole map. Other drivers deal with serializable formats, such as XTM, LTM and AsTMa=.

At this stage, all drivers use a TMRM representation of Topic Maps, instead of adopting TMDM. TMRM exposes a much flatter structure than TMDM does; this is beneficial for the serialization and deserialization process.

5   Knowledge Layer

Once the original resources have their equivalent in Topic Map form, an application can access it. It can do so solely based on the provided ontology, or if that is missing, it can fall back to introspecting the TM content. But this only works sufficiently in those special cases where the one resource contains all relevant information for the application and contains it in exactly that form the application needs and understands. In the case that the relevant information comes from different resources, these maps have to be merged, preferably via the generic TM merging process.

More complex is the case where the application expects the information in a different form as is delivered by the virtualization. An application, for example, may assume that a **person** topic has **author** occurrences whereas the resource has modelled the **author** relationship with dedicated associations.

One may consider adapting the application to the appropriate structure, but this may be a rather expensive process, especially during the maintenance phase. Another option is to write the application very flexibly in the first place and adapt the configuration settings. But this is expensive too, and it may also come with a considerably performance penalty.

An alternative solution for this problem is to let the middleware transform the maps. To control how resources are added and or transformed, a simple algebra has been developed for this purpose. The atoms in this algebra are maps, be they virtual or not. Using the binary, symmetric operator + between maps indicates Topic Map merging, the * indicates a transformation. We assume a processor which will interpret expressions in this algebra as follows.

5.1 Tau Algebra

To address individual maps, to combine them into larger ones and to apply filter transformations, we introduce a simple algebraic language. In the following we only motivate this with a number of examples.

Individual Maps. Quite naturally, we use URIs to address maps so that we can use `file:map.xtm` or the shorter form `map.xtm` to point to a map stored in a file with that name in XTM notation. In any case, the expectation is that the processor will consume the named resource and produce the map therein for further processing.

Another reserved scheme is `null:`; it represents the empty map and is mostly used for formal reasons. More of practical importance are two special maps used for I/O, `io:stdin` and `io:stdout`.

Drivers. It is expected that Tau processors will support several access schemes, so that also remote maps can be addressed. Processors also may support any number of Topic Map notations, not only XTM.

To make this convenient for the user, processors will have to adopt some heuristics to *guess* the provenance of the Topic Map content, such that the correct (network) access is used and the correct import mechanism (notation parser, database connector) is selected.

While this will work in many cases, especially in those where meta data (MIME types or UNIX `magic`, for example) exists, it may fail in others. As an override, users can explicitly name a driver whereby the specifics may depend on implementations. In the following example we use a Perl package namespace to enforce that the map is interpreted as one encoded in AsTMa=:

```
file:map { TM::Materialized::AsTMa }
```

Merging. Combining maps can be done in several ways. One method inherent in the paradigm is *generic merging*. In this process topics in the involved maps are *identified* via their subject addresses or their subject indicators. Once this is done, all characteristics are shuffled together into one combined topic. Applications are also allowed to do *application-specific merging* by using other, possibly external information to decide whether two topics are about the same subject.

In the following we symbolize generic merging with the binary operator + defined between maps, so that we can write

```
file:map1.xtm + file:map2.ltm
```

and expect to receive a combined, merged map at our disposal.

Not only maps can be *added* together; it is also possible to add ontological knowledge to a map, thus overloading the + operator. In effect, all vocabulary and the taxonomy structure in the ontology is merged into the map. If the ontology also contains inference rules, then these will be used when the map is queried or processed otherwise. An example would be a map wrapping the global DNS service:

```
dns:dns.example.org + file:dns.onto
```

Provided the processor supports such a wrapper, it would use the DNS server on dns.example.org for the forward and reverse lookup between IP addresses and host names. The DNS terminology itself is to be found in file:dns.onto.

In any case, we assume the + operation to be commutative and associative. It is worth noting, though, that merging cannot always be materialized, as the example with the DNS wrapper clearly shows. Only if both operands are materializable, the merged result is materializable. Still, it is up to the processor to decide whether the materialization can or should take place when evaluating this subexpression; or whether it is deferred until a later processing (lazy evaluation).

Filtering and Transforming. Maps can also be filtered or, more generally, transformed. One way is to use existing functionality, such as provided by a *statistics* package:

```
file:map.xtm * http://psi...org/tau/statistics/
```

Processors which have a registered implementation for this namespace will take the left operand map to compute statistical information for it. The result of this filter application will be a new map, one containing topics holding the statistical information, such as the number of topics, number of associations, etc.

More flexible transformers can be those written in TMQL. Assuming that in file:services.atm we had listed services running on particular hosts and that in servers.tmql we had formulated a query which would filter out those hosts on our network which have at least one service running, then the expression

```
( services.atm + dns:dns.example.org ) * servers.tmql
```

could render a map containing only those topics representing server hosts.

Ontologies can be interpreted as filters, in that the result is only that part of a map which conforms to the rules and constraints provided in the ontology. To look only for DNS-related information in a map, one would write:

```
file:map.xtm * file:dns.onto
```

Here the result would be a map with only those topics (or associations) which are instances of concepts declared in the file dns.onto. We also expect the resulting map to satisfy any other constraint listed in file:dns.onto.

Input/Output. Obviously, when maps are addressed via URIs they have to be read from the specified resource and have to be deserialized before further processing. After that, the result has to be stored somewhere. For this purpose, the processor understands the concept of *synchronizing* between topic map content stored externally and the internal (memory) representation. At the beginning of an evaluation, content is 'synchronized in', at the end 'synchronized out'.

To make the formalism consistent, we have introduced the symbol > to separate the source and the target in an expression. The simple expression `file:map.xtm` is actually only an abbreviation for the more explicit

```
file:map.xtm > io:stdout
```

It should indicate that — when map content is to be synchronized into the system — the resource `file:map.xtm` should be used; when map content is to be externalized, then the special resource `io:stdout` should be used. What the sink `io:stdout` actually outputs, may depend on the context the application provides.

The mechanism also opens a simple way to convert between topic map notations, such as

```
map.atm > map.xtm
```

or between RDF and Topic Maps

```
http://example.org/me.foaf > me.atm
```

or to create a topic map snapshot of a (wrapped) resource

```
rdb:mysql:host:students > snapshot.xtm
```

Here we exhaustively copied a map extracted from a relational database using a custom-made driver and copied it into an XTM file. If the database wrapper would support the modification of the data in the tables, then

```
null: > rdb:mysql:host:students
```

would get rid of all student information.

UNIX Style. To mimick a more UNIX-like notation we allow the symbol – to be used instead of `io:stdin` and `io:stdout`. Because of the context in which – is used, the processor can always decide whether it is to be interpreted as STDIN, STDOUT or something else the application has specified. Following this scheme, we can input from STDIN and store the map content somewhere

```
- { TM::Materialized::LTM } > file:map.xtm
```

given that the processor knows how to access content from STDIN and that the appropriate LTM driver package is available.

The intention was to mimick UNIX file handling in the sense that `> file.atm` should mean *write the content to the file at the end of the process*. In the same vein, `file:map.atm` indicates that the file should be consumed at the start of the process, but should not be tampered with thereafter. Both these cases may not make much sense within this evaluation model. But the very same expression language can be used also within application code via an API taking care of all necessary input/output mechanics.

In the case a map should be read first, and should finally be written back at the end, then

```
file:map.atm > file:map.atm
```

achieves this. To avoid the noise, we allow to write

```
> file:map.atm >
```

instead.

Avoiding Synchronization. Using the default synchronization behavior — loading at the beginning of evaluation, saving at the end — is not always desirable or possible. When virtualizing of a resource such as the global DNS service, just writing `dns:dns.example.org` would be equivalent with `dns:dns.example.org > -`. Theoretically, this would dump the complete DNS content onto STDOUT (provided the wrapper were cooperative).

The way this is handled, is that drivers are free to deny enumeration; or any other particular access pattern, for that matter.

5.2 Map Spheres

To cope with a larger collection of maps we introduce the additional concept of *map spheres* as a repository of maps whereby each map is addressable with a unique path, such as `/systems/`, `/internet/web/` or `/internet/web/semantic/`. We have reserved one URI scheme, `tm`, to address maps in the map sphere attached to a processor (violating somewhat practice 2.4, *Reuse URI schemes* in the WWW Architecture [11]):

```
tm://internet/web/browsers/
tm://markup/xml/xslt/
```

This process of organizing maps is quite comparable to mounting different file systems into a UNIX file system tree. It makes no difference whether a particular map is virtual or not, which storage technology it is using, or how the namespace is actually formed; the only point is to organize whole maps into a defined hierarchical namespace without merging them. Any such hierarchy itself is completely local and only serves as a convenient way to address whole maps.

Once a map sphere has been created, it can be treated as virtual map. If we would query a sphere for all its top-level topics we would get a list of direct child maps; in the example above these topics would be `internet` and `systems`. Querying for topics under `/internet/` would return only `web` in that map. The hierarchical namespace provides a simple mechanism to *zooming*: At every level of the namespace tree, the embedded maps are represented as topics of a pre-defined type `topicmap`. A child map is then simply attached to its representing topic using reification.

Map spheres can also be distributed over several nodes. For this purpose one map storage technology actually uses the TMIP [4] client/server protocol. If the client is mounted into a map sphere, any access for this path is forwarded to another network node using TMIP. There a TMIP server will receive the access request. Since the server itself is nothing else than an HTTP frontend to a local map sphere, that can either satisfy the request directly or relay it to other nodes.

6   Service Orientation

Syndicates — by their nature — communicate with upstream and downstream partners mostly with headline and news centric protocols, such as HTML over HTTP, RSS and Atom over HTTP or maybe NewsML over SOAP.

On the middleware level we need to avoid any semantic impedance mismatch when exchanging arbitrary knowledge between knowledge syndicating partners. For this purpose a generic TM protocol, such as TMIP [5] or TMRap [9] can be deployed. Both enable bulk transfer of whole maps, of fragments thereof but also support a certain degree of update functionality.

Generally, when designing service protocols a major design choice is whether the protocol follows a convential web services philosophy (using SOAP/WSDL), or whether it adopts a RESTful [8] approach. In the former case, a service description is created which communicating partners can later interpret. The advantage of this explicit description is, that it can be very detailed in the way how and when information flows.

The RESTful approach is more disciplined. It does *not* use an explicit service description but instead views every information on the web as a *resource* [11]. It peruses the generic HTTP methods GET, POST, PUT, DELETE, etc.

In the following we examine the merits of the RESTful TMIP along some typical exchange scenarios.

6.1   Exchanging Maps

To download a complete map from a map sphere a simple HTTP GET request is sufficient:

```
GET /internet/web/ HTTP/1.1
Host: server1.farm.example.org
Accepts: application/xtm+xml
```

The URI `/internet/web/` addresses the map on the service end point. As we are using HTTP 1.1 we are also supposed to send a `Host` header to support virtual hosting. More importantly the protocol supports *content negotiation* whereby the client can request the information from the server in a particular format. This is indicated with the `Accepts` header. It can list all the MIME types which the client is willing to accept. The `application/xtm+xml` is the predefined MIME type for XTM [10], the official XML notation for Topic Maps. Content negotiation allows to use alternative formats, specifically those which are much more concise than XTM.

Should there be no map at the given URI, the server will respond with an HTTP-specific *not found* message. Similar error messages are returned if the server is unwilling to perform the serialization of the map into XTM or if it cannot access its backend store(s). Otherwise the server will respond with

```
HTTP/1.1 200 OK
Server: TMIP server v0.3
Content-Type: application/xtm+xml

<?xml version="1.0"?>
<topicMap.....
```

rendering the map in the requested format.

To create a new map or completely replace an existing one, the HTTP PUT method can be used:

```
PUT /internet/web/ HTTP/1.1
Host: server1.farm.example.org
Content-Type: text/x-astma

google isa search-engine
....
```

Should the map be deserializable and should the privileges of the client be sufficient, then the map is established on the server side:

```
HTTP/1.1 201 Created
Server: TMIP server v0.3
```

6.2   Exchanging Fragments

As the TMDM is quite bulky, the exchange of map fragments, specifically topics, associations and occurrences takes more effort. To retrieve a single topic is quite straightforward, though:

```
GET /internet/web/firefox HTTP/1.1
Host: server1.farm.example.org
Accepts: ...
```

The first part of the address specifies the map within the map sphere, in the above case /internet/web/. Then it continues to detail the topic by its identifier firefox (also subject indicators and subject locators can be used instead). That way the client requests the whole topic, i.e. all names and all occurrences. Also associations can be extracted that way as long as their internal identifier is known (or they have been reified). Which format to use is again under the control of the client.

In many cases, though, clients will not need complete TM structures. A web portal, for instance, might only be interested in particular aspects of a set of topics which satisfy certain criteria. Conceptually, this amounts to *querying* the map on the server side. To make this as much as possible convenient and transparent to the client, it can use a TMQL syntax style which is using path expressions only:

```
GET /internet/web/ //browser / name HTTP/1.1
Accepts: text/xml
```

Like before, first the map inside the map sphere has to be addressed. What follows then is a TMQL path expression, one which first finds all instances of the concept browser in the map and then extracts the names from each of these topics. As names consist not only of the string itself but have type and scope, the TMQL default is to *literalify* the name by simply using the name string only. TMQL allows to suppress this behaviour, if that is unwarranted.

In any case the result is a list of tuples, each tuple containing only a single string. This is — as the client has requested above — serialized into XML:

```
HTTP/1.1 200 OK
Content-Type: application/x-tmql-sequence+xml

<seq xmlns="http://astma..../ns/tmip/1.0/ts/">
  <t><s>Firefox</s></t>
  <t><s>Mozilla Firefox</s></t>
  <t><s>w3m</s></t>
  ....
</seq>
```

This XML format is the only format mandated and defined by TMIP.

For more advanced queries TMQL path expressions will become quite convoluted. In these situations the client may choose to use one of the other available syntactic flavours of TMQL and send a whole query expression inside the message body:

```
GET /internet/web/ HTTP/1.1
Accepts: text/xml
Content-Type: application/x-tmql

return
<browsers>{
    for $browser in // browser
```

```
return
    <browser id="{ $browser }">{
        for $n in $browser / name   return
            <name>{ $n }</name>
    }</browser>
}</browsers>
```

Also this query iterates over all instances of **browser**. In the inner loop it collects all names of one browser topic and wraps it into **<name>** tags. FLWR-style queries like these provide a high degree of control what information is to be returned and also how it is serialized.

6.3   Updating Fragments

Clients cannot only update whole maps but also fragments thereof. Which parts of a map is supposed to be updated can be specified with a TMQL query. The body of the message then contains the update information:

```
PUT /internet/web/ firefox / download HTTP/1.1
Content-Type: application/x-tmql-sequence+xml

<seq>
    <t><s>http://www.spreadfirefox.com/</s></t>
</seq>
```

In the above case the **firefox** topic is going to be modified. This is detailed by the path expression / **download** which selects *all occurrences* of type **download** from that topic. This set of occurrences will be replaced by what the content in HTTP body provides. If the existing occurrences should stay untouched, the method POST (instead of PUT) would have to be used.

The client can also add or replace topics and associations. Again, a path expression will first select those parts in the map affected; the body of the message is then interpreted as new content.

7   Observations and Future Work

To assess the implementation costs of the proposed framework, it was proto-typed in one of the mainstream high-level languages. Also several wrappers have been implemented to stress-test the software architecture, mainly to use the framework as part of simple applications. Most of the work concentrates on the TMQL processor as it is the main workhorse in the middleware: the service protocol TMIP depends on it, the merging wrappers have to use it to analyze relevant ontologies. It is used in transformers and transformer chains and, of course, the application itself.

This highlights the importance of TMQL in our infrastructure and raises questions regarding more optimization techniques for query processors. First ad-hoc results have shown dramatic improvements on practical query times. More work in this area has to follow.

7.1 TMIP Performance

To understand the quantitive performance indicators several benchmarks have been run [4]. The experiments included a client/server exchange of TM fragments using different serialization formats. While it was expected that a large portion of the computational cost is concentrated in evaluating TMQL query expressions, the numbers have indicated that currently at least two magnitudes lie between the exchange process (serialization, transfer, deserialization and dispatch on the server side) and the evaluation of the query. This ratio increased strongly with larger, more realistic topic map sizes.

7.2 Aspect-Orientation

One typical problem middleware has to address is to allow different applications or even one and the same to have a different view on existing resources. In our architecture these *database views* can be quite naturally provided by using different wrappers, each of them is described and associated with its own ontology. It is then a matter of choice (or performance) whether the application is fit to the wrapped resource, or the other way round.

In some cases — which still have to be properly analyzed — the discrepancies between the ontology the wrapper offers and the ontological expectations of the application can be bridged with a transformer. Our expectation is that TMQL can play a strong role, although it can handle only unidirectional mappings between ontologies. This implies that for each of the directions a different TMQL expression will have to be created. This area mandates more research into developing tool support for this process.

Another tangent in this context is how an application is actually experiencing the TM data. One approach is to let the application deal with *topics, occurrences* and *associations* objects. It will then use specialized functions of a TM API to access the information within these topic map items.

These objects can be created either with the TM API itself, or the application may choose to use TMQL as access mechanism. In either case, we do not distinguish between the *instance data* coming from the wrapped resource and the item information inside the *ontology* which describes the interface. Also in both cases the ontological information can be used for inferencing purposes.

Alternatively the application can choose not to receive topic map items but application objects. In this case the middleware will need knowledge how TM information is to be mapped into application-specific objects. Topics of type `person` will then be converted into objects of class `PERSON`, and the other way round, if necessary. This latter functionality is still missing from our infrastructure.

7.3 Change Management

Change management addresses the problem of evolving requirements, of evolving data corpii and of evolving technologies. All of these can be addressed with virtualization, albeit with sometimes considerable cost involved.

Should data be migrated from one storage system into another, then in our framework only the wrapper has to be replaced. The ontology and hence the interface remains constant. Should the application requirements change and the existing virtual map becomes insufficient to provide enough information, then the virtual map can be extended by additional virtual maps via merging.

Change management can also be quite ephemeral, such as in transactions. To solve this within the paradigm of virtual maps, a new virtual map is created for every started transaction. Any access to the map is then routed through the transaction wrapper until the transaction has ended. That will provide to the using application a view which includes all changes within that transaction while referring to the existing content which has not been updated. Other applications will use the original view, so that they cannot experience any updates. When the transaction is committed, the changes in the transactive map are manifested into the wrapped resource.

One severe defect of our abstraction is that with TMQL only read and write access *from the application into the virtual map* is covered. Should one aspect of the application modify information in the underlying resource, there is no way to notify other aspects about these changes.

Non-acknowledgements

Some of this work was conducted inspite the intellectually adverse conditions at Bond University.

References

1. International Organization for Standardization, ISO/IEC 13250, Information technology - SGML applications - Topic Maps (2000)
2. XQuery, W3C candidate recommendation (2006)
3. Barta, R.: Virtual and federated Topic Maps (2004)
4. Barta, R.: TMIP, a RESTful Topic Maps interaction protocol. In: Extreme 2005 Montreal (2005)
5. Barta, R.: TMIP, Topic Map Interaction Protocol 0.3, specification, Technical Report (2005)
6. Cowan, J., Tobin, R.: XML information set, W3C recommendation, 2nd edn (2004)
7. Fensel, D., Hendler, J.A., Lieberman, H., Wahlster, W. (eds.): Spinning the Semantic Web. MIT Press, Cambridge (2003)
8. Fielding, R.: Architectural styles and the design of network-based software architectures, phd (2002)
9. Garshol, L.M.: Tmrap - Topic maps remote access protocol. In: Maicher, L., Park, J. (eds.) TMRA 2005. LNCS (LNAI), vol. 3873. Springer, Heidelberg (2006)

10. Garshol, L.M., Moore, G.: ISO/IEC JTC1/SC34, Topic Maps - XML syntax (January 25, 2004)
11. Jacobs, I., Walsh, N.: WWWA, architecture of the world wide web, vol. 1 (2004)
12. Vdovjak, R., Houben, G.-J.: Rdf-based architecture for semantic integration of heterogeneous information sources. In: Workshop on Information Integration on the Web (2001)

Large Scale Knowledge Representation of Distributed Biomedical Information

Volker Stümpflen, Thorsten Barnickel, and Karamfilka Nenova

MIPS / Institute for Bioinformatics, GSF - National Research Center for Environment and
Health, Ingolstädter Landstrasse 1, D-85764 Neuherberg, Germany
{v.stuempflen,thorsten.barnickel,karamfilka.nenova}@gsf.de

Abstract. Within the last years the Web dramatically influenced biomedical
research. Although it allows for almost instantaneous access to a huge amount
of distributed information the problem how to retrieve useful information still
persist. With semantic technologies (especially Topic Maps) the solution
becomes tangible. We will discuss in this paper concepts and a technical
realization for knowledge representation within the biomedical domain. This
includes not only the semantic access of distributed and heterogeneous
resources based on state-of-the-art enterprise integration technologies (J2EE,
Web Services) but also an approach for Topic Map based views on unstructured
information from scientific publications. We will furthermore present the
implementation of an information portal based on the seamless semantic
integration of ~ 500 genome databases and ~16.000.000 abstracts.

Keywords: Biology, Web Services, Enterprise Information Integration,
Dynamic Topic Map views, textmining, unstructured data.

1 Introduction

Without any doubt the Web has become the most important medium for innumerable
communities to share their knowledge. Today, it serves for example as a mandatory
prerequisite in biological and biomedical[1] research due to it's capability to provide
almost instantaneously access to information of many thousands of involved scientific
communities and changed dramatically the way research is performed today.
However the shortcomings related to the identification of relevant knowledge is well
known and subject of evolving semantic technologies. Beyond the initial hype it
turned out that successful utilization of these technologies is certainly more
demanding than expected at the beginning. We will discuss in the paper conceptual
approaches and technological realizations for large scale knowledge extraction and
representation based on Topic Maps. Although the biomedical domains are well
suited for the inauguration of semantic technologies these approaches are not
exclusively restricted to them.

---

[1] Because of the strong relationship between both areas we will use the term biomedical
throughout the following text.

L. Maicher and L.M. Garshol (Eds.): TMRA 2007, LNAI 4999, pp. 116–127, 2008.
© Springer-Verlag Berlin Heidelberg 2008

1.1 Present Situation

Before going into technological details it is important to bring the initial situation into the mind. Biomedical knowledge and underlying information is typically segregated into three overlapping domains (see Fig. 1). The unfavorable initial situation at the moment is the somewhat fuzzy interference of typically computer generated and structured information and unstructured human knowledge, both in databases and written publications. Due to the variety of different scientific domains the databases are in addition extremely heterogeneous in content and structure.

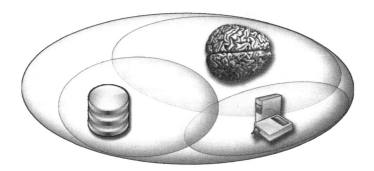

Fig. 1. Overlapping domains of human knowledge, databases and publications. Human knowledge is found both in publications and as free text entries in databases. On the other hand data suited for structured storage in databases may also be found in free text.

Other problems result from the vast amount of freely accessible biomedical data in general (> 1-2 Petabyte) and databases in detail (more than 1000). Even the access mechanisms vary dramatically from downloadable flat-files or direct database access, over web pages up to Web Services.

However in contrast to many other knowledge domains in the web, the biomedical domain is in one important point already very well suited for semantic technologies: the usage of ontologies. Ontologies are inherently related to semantic technologies because they provide the foundation for a computer interpretable description of the meaning of the underlying data. Due to the long term tradition as an empirical science a broad range of controlled vocabulary and ontologies is already used for the description of biological data and collected at obofoundry.org.

1.2 Requirements for Large Scale Knowledge Representation

An adequate way of organizing and storing data is an indispensable prerequisite for an efficient retrieval and subsequent processing of any kind of information. Along general lines, the structure of a data storage unit should resemble the semantic categories and relations humans have in mind when they think of entities of a specific topic. With the tacit understanding that the semantic categories of most humans are similar, this way of data organization facilitates the linking of independently created and maintained data repositories dealing with the same type of entities. In the biological domain different groups and institutions collect and maintain information on genes, proteins and other biological entities independently from each other. Huge amounts of experimentally collected data are stored in relational or object oriented databases and, for the most part, in the form of natural language text. Semantic structuring requires at least the following tasks:

- Development of Topic Maps reflecting the knowledge domains of interest
- Identification of topics, associations and suitable ontology terms in databases
- Semantic extraction of information from free text (Topic Map compatible)

A semantic based structuring and storage of this data is an essential prerequisite for the future linkage of these insulated sources of knowledge. Since related biological information is spread across highly distributed data resources, it is essential to provide powerful techniques for complex searching and retrieval of biological information. Biological data changes continuously and integration of data resources by its replication in huge repositories is even more challenging, since the updating and maintaining mechanisms are rather complex. Hence the migration of existing data into large Topic Maps is not feasible. Instead of this approach, powerful mechanisms capable to integrate internal and external information are required:

- Integration with state-of-the-art enterprise integration technologies. E.g. J2EE or Web Service based
- Intermediate Topic Map Web Services / EJB components for transparent and distributed access to legacy resources
- Dynamic views on Topic Maps generated at runtime

2   Semantic Structuring of Biological Data

Generally, the data stored in databases is already arranged in somehow semantic way, since objects in object oriented databases respectively tables in relational databases usually correspond to real-world subjects, e.g. a protein object or a table containing information related to proteins. Although there are drawbacks like the ambiguity of life science terms and the diversity of data exchange formats [9], the development of suitable semantic structures on top of biomedical databases is more a matter of the particular knowledge domains to be modeled and will go with its details beyond the scope of this paper. Therefore we will focus in the next subsection on the semantic structuring of biomedical text.

2.1   Semantic Structuring of Biomedical Texts: From Texts to Topic Maps

Information contained in natural language text is comparatively unstructured and difficult to interpret by a non-human actor. The amount of biomedical knowledge recorded in texts has been growing tremendously over the last years and the speed of this development is still accelerating [1]. Therefore computer based techniques which are able to extract knowledge from natural language texts, structure it semantically and integrate it to existing knowledge resources are urgently required and text mining systems based on a multitude of different approaches have gained much attention by researchers from academia as well as industry.

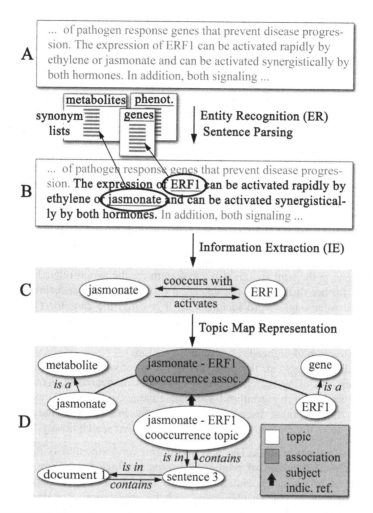

Fig. 2. REBIMET: From Text to Topic Maps: The positions of gene, protein and phenotype names within a document (A) are determined in the *ER* step and sentences containing at least two biological entities are extracted (B). During *IE*, relations between those entities can be derived using cooccurrence or NLP based methods (C). The retrieved relations can be modeled as Topic Map (D: model fragment) using reification to make assertions about associations.

Fig. 2 displays the core principle of our text mining engine REBIMET (Relation Extraction from Biomedical Texts).

REBIMET combines common approaches for a semantic analysis of texts, a storage engine for preprocessed text serving the information for Topic Map representations and an interface for *GeKnowME* (see. Section 3) for the seamless integration with other Topic Map enabled resources:

Synonym Lists and Indexing. The REBIMET system makes use of the large amount of ontologies, controlled vocabularies and databases already available for the biomedical domain to extract lists of biological terms and synonyms needed in the text search step: 20.000 human gene names (15.000 genes) [4], 22.000 metabolite names (13.000 metabolites) [5], 12.000 disease names (4.000 diseases) [6] and over 2.000 tissue terms [7] were collected from various sources, highly ambiguous terms were removed and additional synonyms were added. Disease and tissue terms were subsumed as "phenotypes" as tissue terms often are part of descriptions referring to a disease or distinctive phenotypic feature. In addition, 16 Mio biomedical abstracts were derived from Medline[14] (Dec 2006), the most important literature database in the biomedical domain and added to an index in order to enable an efficient search of the extracted terms in this collection of articles.

Entity Recognition (ER). Entity Recognition in biology refers to the task of recognizing entities like genes, proteins, metabolites or diseases in texts and linking those text sections to entries in external data resources. The recognition of biological entities in text is a very complex process. A major problem is the ambiguity of terms within and between biological domains and between biological descriptors and common English terms. For example, the gene name "art" is the official name for the aristatarsia gene in the fruit fly but also a synonym for the agouti related protein gene in mouse. A further difficulty is the detection of descriptions consisting of multiple words instead of a single word identifier as it is often the case for phenotypes or diseases. Figure 2 B shows a document which has become semantically "annotated" by identifying entities of a certain biological category and is now ready for the automated extraction of higher-level semantic relations.

In the REBIMET system, the positions of all synonyms were determined in the indexed documents using search algorithms of differing degree of fuzziness (a case sensitive and strict search algorithm for the gene names, a fuzzy and case insensitive algorithm for the compound disease names) and stored in a database. Index generation and searching were done based on the Jakarta Lucene text search library.

Information Extraction (IE). The aim of information extraction is the automated retrieval of a predefined set of relations between (biological) entities from natural language text. A relation of potential interest to a biologist is for example the causal relationship between a certain phenotype and the application of a drug. This task crucially depends on the results of "lower" processing steps like ER, sentence splitting and often part of speech (POS) tagging and syntactical sentence parsing. While approaches based on pattern matching have only restricted generalisability due to their closeness to the lexical level of the text, full parsing approaches using the syntactical structure of a sentence to extract meaningful relations between entities often are computationally expensive and not always applicable to huge literature resources like Medline. However, the speed of modern CPUs and the quality of the syntactical parsers (for a benchmarking test of the most prominent parsers in the biological domain see [2]) have increased significantly over the last years. A comparatively new approach called "semantic role labeling" (SRL) extracts subject – predicate – argument structures (SPAs) from sentences and has proven to be useful for the extraction of relations from natural language texts also in the biological domain [3].

REBIMET determines biological entities cooccurring within one sentence to extract semantically related entities, a simple but widespread approach which does not rely on complex and often time consuming NLP preprocessing steps and is therefore frequently used for processing large amounts of text. This approach assumes that cooccurring entities usually are also semantically related. However, as the type of the extracted relation remains unknown in the cooccurrence based approach, REBIMET uses in addition SRL (ASSERT tool [8]) to extract SPAs from sentences containing cooccurrences. These SPAs can be used to extract relations of a certain, predefined type, e.g. "activation" or "inhibition". Figure 2 C depicts the two relation types between the metabolite "jasmonate" and the gene "ERF1" that can be extracted using cooccurrence based IE (upper double-headed arrow) or SRL (directed arrow below).

Modeling Semantic Relations. The effect of the extraction of cooccurring or otherwise semantically connected biological entities is a semantic structuring of the text documents by providing a "navigation framework" that can be modeled as Topic Map (Figure 2 D): Biological entities (topics) are connected to each other via a certain type of semantic relation, e.g. "cooccurrence" or a typed relation like "activation", which are represented as association. As in the Topic Map standard only topics are allowed to have occurrences, reification is used in order to make assertions about associations. In the case of cooccurring entities, reification connects a specific "cooccurrence" association to sentences mentioning both biological entities. If, in addition, a typed semantic relation like "activation" has been detected for this pair of entities, this "activation" association can be connected to all SPAs from which this relation has been extracted without having to change the underlying model.

3 Semantic Integration of Large Scale Biological Data

Common approaches to integrate distributed resources are based on component and/or service oriented multi-tier architectures. The basic idea behind is the segregation of software systems in weakly connected layers performing distinct tasks. The programming platform Java EE (Java Platform, Enterprise Edition) provides various technologies, such as JDBC (Java Database Connectivity), EJB (Enterprise Java Beans), Web Services, XML, etc., for developing such software architectures for solving numerous technical problems by retrieval and exchange of highly distributed data.

However, a limitation of these integration technologies is that they are semantically weak, i.e. the data can be transmitted among diverse applications but unfortunately the data content can not be controlled whether it is in the right context or whether it has the right meaning. Therefore, it is quite significant that the biological data is semantically structured as mentioned in the previous section. We developed a generic and easily extendable framework called *GeKnowME* (Generic Knowledge Modeling Environment) based on the Java EE integration technologies in combination with the key concepts of the Topic Maps standard. By applying this framework it is possible to integrate distributed resources by providing semantics dynamically to the underlying data sources for better and more effective information exploration.

3.1 Concepts

The inspiration for the development of the *GeKnowME* framework was to increase the knowledge transparency in life science by overlaying an independent semantic

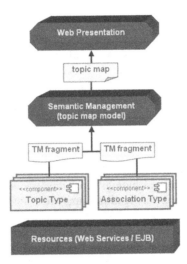

Fig. 3. Dynamic Topic Map Generation. An independent semantic layer is placed on top of any arbitrary data sources. The semantic is appended by *Topic Type* and *Association Type* components by generating TM fragments dynamically, which are combined together to a resultant topic map and presented to the user.

layer on top of any arbitrary data resource. The biological information should be interconnected within a huge knowledge network, which can be navigable in any direction and easily partitioned in sub-networks. To achieve this goal carefully selected technical approaches were essential. An overview of the procedure how this can be realized is shown in Fig. 3.

Since the existing biological data is highly distributed and in addition frequently updated, the integration should be based not on data replication, but on dynamical data retrieval via Web Services or EJBs. Thereupon, a semantic layer as topic map model can be placed to provide semantic integration. Basically, every topic map is composed of topics and associations. Therefore for all existing topic types and association types defined within the topic map model a software component is developed to interpret the related information in a semantic way in the form of a topic map fragment. All associated fragments can be then easily assembled as a resultant topic map and represented to the user.

3.2 Implementation

By the design of the system architecture of the *GeKnowME* framework the multi-tier approach was followed, whereby each single layer performs a distinct task. In fact, the system consists of five different tiers as shown in Fig. 4 and each one represents another technological step in order to solve one or more integration challenges. In addition, the framework is based on the implementation of encapsulated and reusable components that can be deployed and run as EJBs or WSs under Java EE capable application servers like Glassfish or JBoss. A component may hide any arbitrary complexity and is able to communicate in distributed systems with other components

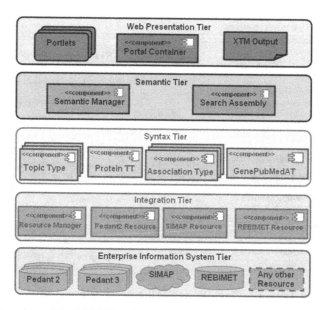

Fig. 4. An Overview of the Multi-Tier System Architecture. The system is divided into five layers each representing another step in the semantic integration.

not only within the own physical environment but also over the network. In the following the main purpose of every single tier will be presented with its corresponding components.

Enterprise Information System Tier: The lowest level in the system architecture represents available resources that contain valuable biological data, which can be referenced within the topic map models. Any kind of resource can be found within this layer: rational databases, applications, Web Services, etc. The main purpose of this tier is to provide the searched data for the upper tiers. No modifications have to be done to the existing resources, which allows the business and presentation logic to be completely independent of taking care of all updating and maintaining problems.

Integration Tier: The integration tier, as the name states, is responsible for the integration of all resources available in the enterprise information system tier. In order to provide information integration for each available resource a concrete component *ConcreteResource* has to be implemented which is in charge of not only establishing and maintaining the connection to the underlying resource but also of executing syntax specific queries and forwarding results. For example the *Pedant2Resource* is responsible for the Pedant2 database. All these Resource components implement the same interface *Resource* as shown in Fig. 5.

Additionally, a component called *ResourceManager* controls all available resources and also takes care of the invocation of the right *Resource* components. In case that a new resource has to be integrated into the system, a corresponding new *Resource* component has to be implemented and registered by the *ResourceManager*. It is not necessary that all *Resource* components have to run under the same Java EE

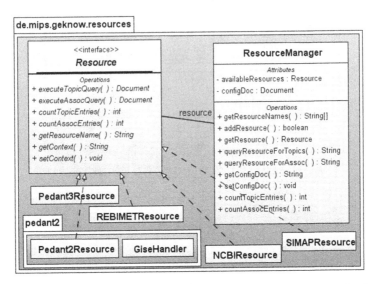

Fig. 5. Integration Tier UML Class Diagram. All concrete resources have to implement the interface *Resource* . All available resources are managed by the *ResourceManager*.

application server. This integration approach allows a very flexible and highly extendable way of information integration, whereby every *Resource* component takes care of all resource specific access procedures.

Syntax Tier: The syntax tier represents an interface between the integration and semantic tier. It is responsible for the interpretation of the searched and retrieved information in the correct way. Some of the Topic Maps key concepts are implemented in this layer. The main idea behind the syntax tier is that every topic map can be defined by the types of all composing topics and association, or in other words by certain concepts and the relations between these concepts. Therefore, the syntax tier consists of two main generic component types: *TopicTypes* and *AssociationTypes*, providing defined interfaces for overall functionality (see Fig. 6).

The implementation of concrete *TopicType* components allows the transformation of information for a certain subject type into a topic maps related one (e.g. topic type "protein"). Each concrete *TopicType* component integrates information related to that subject like existing names, characteristics, identifiers, etc. and provides this information as a single topic map fragment, with the advantage that this information is semantically correct. Correspondingly, the implementation of concrete *AssociationType* components provides the necessary semantic information about the relationship between topics. Depending on the representation of the certain knowledge domains, as many as needed *TopicType* and *AssociationType* components can be developed.

Semantic Tier: The components within the semantic tier were developed with the main purpose to structure and organize the distributed information on a semantic level. The conceptual model of the semantic integration is configurable in this integration step. A component called *SemanticManager* takes care of how the existing

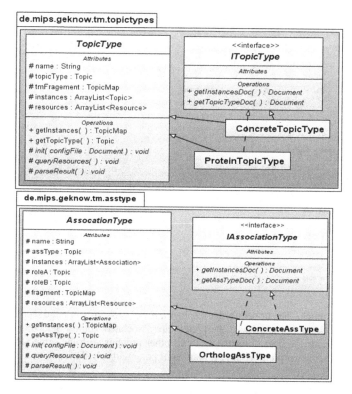

Fig. 6. Syntax Tier UML Class Diagram. Two different kinds of components exist is the syntax tier – *TopicType* and *AssociationType* with their concrete implementations.

topic and association types within the topic map model interact with each other. It has a "dispatcher" role during the composition of the semantic network. Additionally, the component *SearchAssembly* is responsible for the final assembly of the topic map fragments delivered by the invoked topic and association type components. It applies strict predefined merging rules for the construction of a valid topic map containing the union of all available topics and associations connected directly to the searched subject. This approach allows the seamless combination of partially overlapping information of diverse life science domains, because the underlying information is semantically correct defined.

Web Presentation Tier: The final level of the implementation is the web presentation tier, where the users can interact with the system by querying the available resources and intuitively navigating through the related information within the semantic network. The graphical user interface is based on a web portal technology (Liferay: www. liferay.com) , in which portal pages are composed of web components called *Portlets* that process requests and generate dynamic HTML content. In general, within the *GeKnowME* framework we deliver at the end of each implemented component by default XML documents, whereas the components within the syntax and semantic tier

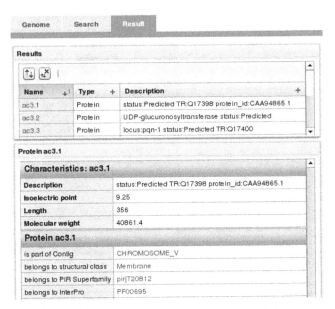

Fig. 7. Portlet example. The upper portlet displays a list of proteins while the second one shows specific results. Certain properties of the protein are associated Topics. Selecting one of these properties will guide the user immediately to all other proteins with the same property, hence allowing him to see the protein within the context of potentially similar ones.

work with valid XTM documents. These generated documents can be processed by predefined portlets (Fig. 7) and represented to the user.

4   Conclusion

Topic Maps provide a straightforward way for the semantic structuring of biomedical data and therefore providing the foundation for extraction and representation of knowledge within the scope of specific interests. In combination with existing ontologies we showed that even the extraction of semantic association of biomedical terms in literature with textmining methods and their integration into Topic Maps become feasible. Furthermore we showed that state-of-the-art enterprise integration technologies (EJBs / Web Services) can be exploited to migrate dynamically information from legacy systems into Topic Maps. Based on a generic implementation we defined Topic Map enabled EJBs / Web Services reducing the mandatory steps down to the implementation of basically two interfaces. In combination with a suitable portal technology we were able to develop a biomedical knowledge representation portal based on the seamless semantic integration of more than 500 genome databases in combination with associated information within literature based on >15.000.000 abstracts.

Acknowledgments. We gratefully acknowledge the 'Impuls- und Vernetzungsfonds' of the Helmholtz-Gesellschaft for funding and Richard Gregory for invaluable discussions.

References

1. Jensen, L.J., Saric, J., Bork, P.: Literature mining for the biologist: From information retrieval to biological discovery. Nat. Rev. Genet. 7, 119–129 (2006)
2. Clegg, A.B., Shepherd, A.J.: Benchmarking natural-language parsers for biological applications using dependency graphs. BMC Bioinformatics 8(24) (2007)
3. Kogan, Y., Collier, N., Pakhomov, S., Krauthammer, M.: Towards semantic role labeling & IE in the medical literature. In: AMIA Annu. Symp. Proc., pp. 410–414 (2005)
4. Pruitt, K.D., Tatusova, T., Maglott, D.R.: NCBI Reference Sequence (RefSeq): A curated non-redundant sequence database of genomes, transcripts and proteins. Nucleic Acids Res., D501–D504 (2005)
5. Kanehisa, M., Goto, S., Hattori, M., Aoki-Kinoshita, K.F., Itoh, M., Kawashima, S., Katayama, T., Araki, M., Hirakawa, M.: From genomics to chemical genomics: New developments in KEGG. Nucleic Acids Res., D354–D357 (2006)
6. McKusick-Nathans Institute of Genetic Medicine, Johns Hopkins University (Baltimore, MD) and National Center for Biotechnology Information, National Library of Medicine (Bethesda, MD): Online Mendelian Inheritance in Man, OMIM (TM). Dez (2006), http://www.ncbi.nlm.nih.gov/omim/
7. Schomburg, I., Chang, A., Ebeling, C., Gremse, M., Heldt, C., Huhn, G., Schomburg, D.: BRENDA, the enzyme database: Updates and major new developments. Nucleic Acids Res., D431–D433 (2004)
8. Pradhan, S., Ward, W., Hacioglu, K., Martin, J., Jurafsky, D.: Shallow Semantic Parsing using Support Vector Machines. In: Proc. of the Human Language Technology Conference/North American chapter of the Association for Computational Linguistics annual meeting (HLT/NAACL-2004), Boston, MA (2004)
9. Stein, L.D.: Integrating biological databases. Nat. Rev. Genet. 4, 337–345 (2003)
10. Cheung, K.H., Yip, K.Y., Smith, A., Deknikker, R., Masiar, A., Gerstein, M.: YeastHub: A semantic web use case for integrating data in the life sciences domain. Bioinformatics 21(suppl 1), 85–96 (2005)
11. Merali, Z., Giles, J.: Databases in peril. Nature 435, 1010-1011 (2005)
12. Topicmaps.Org Authoring Group. XTM Specification (2001)
13. Biezunski, M., Bryan, M., Newcomb, S.R.: ISO/IEC 13250:2000 Topic Maps: Information Technology – Document Description and Markup Languages (2004), http://www.y12.doe.gov/sgml/sc34/document/0129.pdf
14. MEDLINE® (Medical Literature Analysis and Retrieval System Online): NLM's premier bibliographic database. U.S. National Library of Medicine, 8600 Rockville Pike, Bethesda, MD 20894, http://www.ncbi.nlm.nih.gov/sites/entrez?db=PubMed

Versioning of Topic Map Templates and Scalability

Markus Ueberall and Oswald Drobnik

Telematics Group, Institute of Computer Science,
J. W. Goethe-University, D-60054 Frankfurt/Main, Germany
{ueberall,drobnik}@tm.informatik.uni-frankfurt.de

Abstract. Major issues for development environments are version management and scalability support. In this contribution, we discuss design guidelines for building a scalable, Topic Maps based prototype for software development process support, based on our work in [11,12]. Particular attention is paid to the versioning of both Topic Map Templates, i.e., patterns used for making multiple copies of a set of topic map objects, and their instances.

In combination with faceted classification of the resulting hierarchical structures and a generic meta process model, it is shown that the intrinsic traits of Topic Maps can be used to reduce the complexity of the management of large topic maps, e.g., by filtering.

1 Introduction

In the context of software development processes, participants face concepts from multiple domains and may suffer from information overflow caused by version management problems. Ever-ongoing development processes and short going-live-cycles demand light-weight tools to capture and comment information flows as well as interconnections or dependencies in order to rapidly plan and document the evolution of underlying concepts.

Insufficient communication and traceability deficits between participants is another central issue regarding development processes. In particular, information loss results from the inability to synchronise requirements and their implementations in a timely fashion. The problem with "just-in-time" documentation is the adherent effort, especially if third-party systems are involved. What is required is a light-weight, fast means for creation, aggregation and modification of information which complements rather and replaces existing tools.

In the following, an approach is presented which uses Topic Map Templates as a basis for versioning on a conceptual level. Section 2 introduces Topic Map Templates and their relevance for structuring and versioning. Section 3 addresses their use for software development. In section 4, a prototype is proposed and its influence on scalability is discussed. An outlook on further work is given in section 5.

L. Maicher and L.M. Garshol (Eds.): TMRA 2007, LNAI 4999, pp. 128–139, 2008.

2   Topic Map Templates and Versioning

In this section, our notion of *(Nested) Topic Map Templates* is introduced, followed by a discussion of their structuring and versioning using well-known basic metadata standards, ontologies and Topic Map Design Patterns [4].

2.1   Templates and Constraints

Topic Map Templates have been introduced in [1] and extended in [12] to improve the expressiveness of concept definitions for modelling purposes. In this context, a Template consists of any combination of related individual topic map objects, i.e., (typed) topics, associations, and occurrences. It is used for instantiations of concepts as identifiable objects. Part of the definition of a Template is a human-readable description of its underlying semantics.

Nested Templates, as specified in [12], are Templates which include references to other Templates. They may be used to derive new definitions by extensions and/or combinations of existing (basic) concepts. This is similar to the principle of DITA[1] which states that composite concepts should always build on existing "meta-ontologies". However, Topic Map Templates offer semantically richer modelling means, e.g., typed links.

Templates may contain usage restrictions in form of constraints, e.g., cardinality constraints [10] which specify how many instances of occurrences or roles within an association are allowed. Using LTM notation [16], an example for the definition of Topic Map Templates and constraints looks as follows:

```
[person : topic-type = "Person"]
[age : occurrence-type = "Age" = "Person's Age" /person ~toc1]
[toc1 : topic-occurrence-constraint =
     "Topic Occurrence Constraint Label for Occurrence 'Age'"]
{toc1, max-cardinality, [[1]]}
```

The inclusion of all constraints in the topic map instead of using external schema definitions has some important advantages regarding both version management and user interface support as addressed in the following sections.

Scope is another important type of constraint. Cerny [2] recommends that new concepts should be classified ad-hoc during their first use. This correlates with the assignment of basic role constraints for individual topic map objects as, e.g., used within the NCPL [3]: Once a topic is typecast as, e.g., an `occurrence-type`, it is impossible to use it as a `topic-type` at the same time. In addition, the `scope` operator is commonly used for context-sensitive labeling of association roles, cf. `Person's age` vs. `age`.

Alternatively, scope can also be modelled using view definitions, i.e., filter criteria [11], which can be expressed as "in-line queries" in analogy to cardinalitites [12].

---

[1] Darwin Information Typing Architecture, `http://dita.xml.org`

2.2 Structuring of Templates

Hierarchical structuring is necessary in order to simplify the application logic
needed to support Nested Templates. In addition to the basic `supertype-sub-`
`type` relation, which is part of the TMDM specification and therefore already
supported by various Topic Maps applications, more flexible concepts have been
introduced by various authors [4,3,10,13].

In particular, we use the following classification concepts referenced by the
respective URIs/Prefixes:

```
#PREFIX tmhd @"http://www.techquila.com/psi/hierarchy/#"
#PREFIX tmtd @"http://www.techquila.com/psi/thesaurus/#"
#PREFIX tmfd
    @"http://www.techquila.com/psi/faceted-classification/#"
```

The first two references contain concepts that are sufficient to model simple hierarchies, however they are not flexible enough on their own. The third reference
defines a *faceted classification* system, which provides the assignment of multiple
classifications to an object, enabling the classifications to be ordered in multiple
ways rather than in a single, pre-determined, taxonomic order [14].

These concepts are combined with guidelines of [4,3,10], e.g., "avoidance of
grouping types", "reification of data structures", and "use of dedicated (role)
types", to specify nested templates as well as their respective instances as follows:

```
[all-templates : tmfd:facet = "Set of all Templates"]
[template-facet : tmfd:facet = "Template Facet"]
tmfd:facet-has-hierarchy-type(template-facet : tmfd:facet,
    tmfd:subcategory-supercategory : tmfd:facet-hierarchy-type)
tmfd:facet-has-root(template-facet : tmfd:facet,
    all-templates : tmfd:facet-root)

// taken from the UML Superstructure specification, cf. [12]
[actor : template-class = "Actor"]
tmfd:subcategory-supercategory(actor : tmhd:subcategory,
    all-templates : tmhd:supercategory)
tmfd:subcategory-supercategory(person : tmhd:subcategory,
    actor : tmhd:supercategory)
tmfd:subcategory-supercategory(legal-entity :
    tmhd:subcategory, actor : tmhd:supercategory)

tmtd:part-whole(actor : tmtd:whole, role : tmtd:part)
tmtd:part-whole(person : tmtd:whole, age : tmtd:part)
```

This example illustrates how different types of hierarchical relations can be used
to separate attributes from categories.

2.3   Versioning

Versioning is conceived as the management of multiple revisions of the "same" unit of information, i.e., in our context, a template or one of its instances.

Any object which can be referenced should automatically be annotated with a version identifier (id) by the system. In a centralised environment, it is feasible to assign consecutive version numbers; however, there are reasons to do without and, e.g., use a CRC checksum–if users need descriptive revisions, they are always able to tag/annotate each object as with "traditional" revision management systems.

Aside from the version id itself, especially in a collaborative, distributed scenario, a timestamp and means to attribute changes to its author (and his intentions), are needed, too:

```
// <occurrence/association> ~object-ref
// dc: cf. dublincore.org/documents/dces (Dublin Core)
// skos: cf. w3.org/2004/02/skos/ (SKOS)
{object-ref, dc:creation-date, [[2007-06-04T2359:59+01:00]]}
{object-ref, dc:version,  [[version-id]]}
{object-ref, dc:author, [[user-id]]}
{object-ref, skos:changeNote, [[description]]}
...
```

If one version (i.e., data attached to `object-ref`) of a composite template or instance is replaced by another one because a constituent object changed, thereby affecting the semantics of the whole construct, this does not necessarily mean that all (other) associated objects have to be revised, too (cf. fig. 2). All existing objects can be reused which is comparable to the immutable String objects in Java. It is sufficient to mark a version as "outdated" using, e.g., one of the following associations:

```
is-replaced-by(object-ref₁ :old-obj, object-ref₂ :new-obh)
is-deprecated(object-ref₃ :obj)
is-deleted(object-ref₄ :obj)
...
```

The use of the aforementioned versioning definitions provides a simple means to log all kinds of changes which is a crucial prerequisite for traceability. As a side effect, it also maps the complete version history of all managed concepts to a directed acyclic graph (DAG). Such a graph is needed to support distributed development environments. *Decentralised version management* is very complex and depends on a code integrator who merges versions from different repositories in contrast to centralised version management where any priviledged user is able to merge different versions [20].

Our approach of versioning Topic Map Templates seems to be a promising approach to simplify decentralised version management on a conceptual level.

3   Template-Supported Software Development

In this section, the objectives and underlying concepts of the design phase of software development processes are addressed on an abstract level in order to facilitate the understanding of how version management can alleviate re-engineering efforts caused by changing functional or non-functional requirements.

3.1   The Design Phase

The objective of the design phase is to identify and analyse all requirements in order to define the software architecture and the interfaces between individual components. For example, both the business and use cases given by the stakeholders, which basically model the system's internal and external interfaces, may contain the following implicitly or explicitly stated requirements describing both functional and non-functional system qualities:

```
{req103, requirement, [[flight bookings may be cancelled]]}
{req205, requirement, [[online booking system must be
    available at any time]]}
```

To support the implementation phase, a designer has to annotate these requirements for the programmer. In other words, "incoming" requirements from the application domain will essentially be mapped to "outgoing" design-oriented requirements along with needed templates, i.e., the designer acts as a translator or mediator, respectively. The above set of requirements may be extended, e.g., as follows:

```
{req103a, requirement, [[data management must include customer
    and flight related information]]}
{req205a, requirement, [[all components of the booking system
    have to be redundant]]}
```

Design-oriented templates include descriptions of design patterns or architectural patterns, i.e., general reusable solutions to commonly occuring problems in software design [9]. The following example outlines the definition of the "singleton" design pattern and how to describe method signatures the implementation has to adhere to:

```
[singleton : design-pattern = "Singleton"]  ~concept1234
// dc: dublincore.org/documents/dces (Dublin Core)
{singleton, dc:date, [[1995]]}
{singleton, dc:creator, [[Gamma, Helm, Johnson, Vlissides]]}
{singleton, dc:description, [[The Singleton pattern ensures that
    only a single instance of an object exists...]]}
// odol: www-ist.massey.ac.nz/wop/... (Web of Patterns Project)
{constructor, odol:declaration, [[private]]}
{getInstance, odol:declaration, [[public]]}
```

```
tmfd:subcategory-supercategory(design-pattern : tmhd:subcategory,
    all-templates : tmhd:supercategory)
tmtd:part-whole(singleton : tmtd:whole, constructor : tmtd:part)
tmtd:part-whole(singleton : tmtd:whole, getInstance : tmtd:part)
```

In this example, an implicit *subcategory–supercategory* relationship is used, i.e., a singleton *is-a* design-pattern which in turn *is-a* template class. However, the required application specific logic complicates the re-use of this classification hierarchy with other applications, so we do not recommend it. In order to alleviate the problem, the application should make the relationship explicit by adding the 'missing' association (the same goes for the method signature information).

Also, the listing shows how to freely annotate a concept–in this case, a design pattern formulated in [9]–with Dublin Core metadata. Independently from this, the reification (~concept1234) should be added automatically in order to add system-specific metadata as discussed in the previous section.

3.2 Development Phases and Versioning

Large software projects are usually organised in several groups which can work in parallel but have to synchronise depending on the development progress.

In the following, we focus on the requirements analysis, design, and implementation development phases and on the transitions within and between these phases using the flight booking scenario of section 3.1 as depicted in fig. 1.

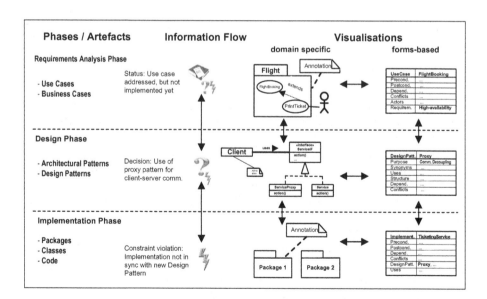

Fig. 1. Motivating Software Development scenario depicting associated phase-specific concepts and information flow as described in subsection 3.2

The availability requirement `req205` stated by the stakeholders leads to a system with redundant components to support load balancing and fault tolerance (see `req205a`). From the software point of view, the direct coupling between clients and services has to be substituted by an indirect coupling, e.g., on the basis of the proxy design pattern (see fig. 2).

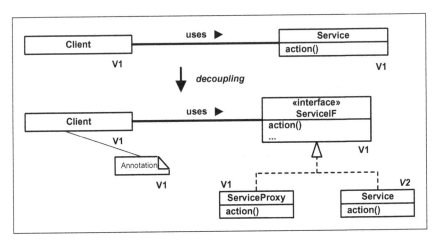

Fig. 2. Introducing the proxy design pattern: The new interface of concept (class) `Service` demands a new revision; concept `client`, on the other hand, is unaffected by a comment as long as that does not imply a change of semantics

This design decision has to be communicated to both stakeholders and programmers in a suitable way: Stakeholders are informed that a system architecture has been projected with respect to their original requirement, programmers are notified to modify the code related to the affected interfaces.

The resulting changes in the associated topic map can be substantial, because all related concepts may have to be changed and, in this case, the proxy pattern as depicted in fig. 2 is to be supplemented–depending on the context, different other patterns adressing, e.g., logging and load-balancing, usually complement it. In practice, this may lead to the revision of all modules that encapsulate the new functionality. Each change means that the respective component gets a new version identifier and has to be revalidated.

The programmers have to document their code changes using templates reflecting the interfaces. In this way, designers and stakeholders can trace the development progress.

Since versioning is done on conceptual level, only changes in semantics should lead to version increments. Fig. 2 demonstrates this for the `Service` class which gets a newly introduced interface. The `client` class on the other hand is unchanged, because neither the added annotation nor the interfaces changes its behaviour. Still, users that view the `client` class will be notified that related oncepts have changed.

3.3   Meta Process Model and Traceability

As shown in the previous subsection, changes in requirements may cause subsequent interactions between all development groups. Therefore, a process model is needed to control the information flow between the groups to make the phase-specific concepts consistent or to resolve conflicts.

A simple, light-weight generic meta process model which can be used to tailor workflows as well as associated roles which encapsulate access control mechanisms has been presented in [11].

Two so-called "summarisation" and "exploration" subphases embrace each complete development step, i.e., after a participant decides to commit his changes, in order to support reconciliation [21]. During the "summarisation" phase, the user has to comment his own changes before they are committed to the repository. During the subsequent "exploration" phase, every other participant who uses concepts that are affected by those changes has to decide whether to accept or reject them. In the latter case, the author of the changes is notified about the rejection and may revise them.

To resolve conflicts from possible misunderstandings regarding concepts of other users due to different domain knowledge, both computer-driven matchings [17] of concepts as well as computer-aided reconciliation [18] can be used.

Implications of changes for other domains/participants might not be obvious right away, therefore versioning is an essential prerequisite for both documentation and configuration management: It allows to trace all changes to concepts and underlying earlier decisions on one hand, and allows to, e.g., experimentally combine different revisions of related concepts on the other hand.

4   Tools and Scalability

In the following, we give some technical details regarding our prototype and its interplay with existing tools. Based on the fundamentals of the previous sections, we discuss scalability.

4.1   Eclipse-Based Prototype

The prototype has to support the management of concepts from all phases of software development as shown in Fig. 1. It should complement existing development environments in such a way that all development groups obtain consistent views on information and workflows. It has to offer a coherent user interface which allows users to freely create or annotate concepts using familiar representations, e.g., designers/programmers may use UML diagrams, stakeholders may enter forms similar to CRC cards and work with mindmaps. Moreover, it should be easy to use, light-weight, and build on freely available tools.

To meet this criteria, our prototype has been based on the Eclipse framework for several reasons: Eclipse is the de-facto standard Java development environment and offers a powerful plug-in architecture. A large number of available libraries, e.g., already support the synchronisation of program code and graphical models.

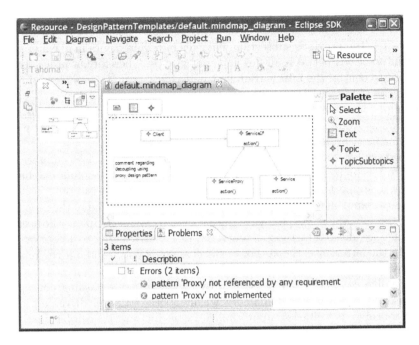

Fig. 3. A snapshot of the Eclipse-based prototype's graphical interface (based on the GMF mindmap example application) showing composite concepts and resulting constraint violations

Because the prototype is implemented as an Eclipse plug-in, it can be combined directly or indirectly with other plug-ins using so-called extension points.

The internal representation of concepts as topic map objects is accessed via the TMAPI[2] used by a number of open source applications, thereby enabling the use of freely available Topic Maps engines, e.g., tinyTIM[3]. Due to performance reasons, we implemented our own engine using the H2 Java database[4]. Since our prototype exploits the ability of Java to dynamically load application logic at runtime, different Topic Maps query and constraint engines can be integrated on demand (cf. [11,12]).

An example of the prototype's interface–which is currently still under development–is shown in fig. 3. It illustrates the graphical representation of concepts from our example scenario on the preceding section. The snapshot shows the change of the design caused by the introduction of the proxy pattern and the resulting constraint violations stating that this pattern has not yet been implemented.

The prototype interacts with the user in correspondence with the meta process model. Predefined roles, e.g., stakeholder, designer, or programmer, control both the visualisation of and access to domain-specific concepts. While a user cannot

---

[2] http://www.tmapi.org
[3] http://tinytim.sourceforge.net/
[4] http://www.h2database.com

modify concepts which do not belong to his responsibility as defined by the role, it is always possible to freely annotate them using a basic set of domain-independent concepts and relations [19]. In this way, e.g., questions regarding the use of templates can be forwarded to their respective authors. In order to support the instantiation of templates, the prototype supports a forms-based input dialog as shown on the right in fig. 1.

Since all user actions are logged, modifications to concepts are versioned automatically as well, which enforces traceability [12]. Unless the user explicitly commits his changes, they are not visible to others, though the affected concepts are marked as being modified.

4.2   Scalability

The application of our method to real software projects implies huge amounts of data. Therefore, scalability support is of major importance. Since our approach exploits the known advantages of Topic Maps based representations and the hierarchical structuring by means of facets, it is possible to navigate through the data in a very systematic way on different conceptual levels.

Querying and filtering [12] enables the user to obtain an abstract view on domain- and phase-specific issues by distinguishing different viewpoints which may be visualised adequately. E.g., in our proxy example, the different viewpoints and objectives depend on the users's role as follows: Stakeholder–use cases/high-availability requirement, designer–system architecture/proxy design pattern, programmer–implementation/decoupling of interfaces.

Moreover, all different views are interlinked in such a way that the associated concepts can be kept consistent. Detected conflicts are reported to all users.

To reduce the complexity of communication between collaborating users on a technical level, a variant of the algorithms presented by Ahmed [5] and Schwotzer [7] can be used to identify topic map objects (forming a fragment) that need to be exchanged in order to broadcast a certain action, e.g., for reconciliation.

This demonstrates the suitability of Topic Maps as a lightweight representation for team collaboration.

5   Conclusion and Outlook

In this contribution, we provided technical details related to the use of Topic Map Templates in modelling applications, especially focusing on the handling of multiple versions of concepts.

Using faceted classification, it is possible to both limit the number of objects that have to be exchanged in collaborative environments and at the same time support navigation between different concept domains.

Based on lessons learned from the ongoing implementation and similar approaches from others [10,3], we conclude that Topic Maps are indeed suited to represent models and instances of concepts from different domains in a lightweight fashion.

Currently, the prototype has to be evaluated with regard to more complex development scenarios. Also, the combination with other Eclipse plug-ins has to be investigated: e.g., the "WebOfPatterns" project[5] provides a design pattern scanner whereas the "KeY" project[6] focuses on formal software specification and verification support.

For coping with ontology changes or mergers, the functionality of the meta process model subphases mentioned in section 3.3 could be enhanced by utilising the algorithms and strategies in [17,18].

Last but not least, the underlying operations of a decentralised version management system are very complex[7], and, as of this writing, are not supported by the prototype presented in subsection 4.1 yet.

References

1. Rath, H.H., Pepper, S.: Topic Maps: Introduction and Allegro. In: Proc. Markup Technologies (1999),
   http://www.ontopia.net/topicmaps/materials/allegro.pdf
2. Cerny, R.: Topincs Wiki: A Topic Maps Wiki. In: Maicher, L., Garshol, L.M. (eds.) Scaling Topic Maps. LNCS (LNAI), vol. 4999, pp. 1–9. Springer, Heidelberg (2007)
3. Ahmed, K.: Design Patterns and Practices using Topic Maps. In: Proc. TM2007,
   www.topicmaps.com/tm2007/tutorials/TmDesignPatternsHandOuts.pdf
4. Ahmed, K.: Beyond PSIs: Topic Map Design Patterns. In: Proc. Extreme Markup Languages (2003), http://www.idealliance.org/papers/extreme03/
   html/2003/Ahmed01/EML2003Ahmed01.html
5. Ahmed, K.: TMShare - Topic Map Fragment Exchange in a Peer-to-Peer Application. In: Proc. XML Europe 2003, http://www.idealliance.org/papers/
   dx_xmle03/papers/02-03-03/02-03-03.html
6. Ahmed, K., Moore, G.: Topic Map Relational Query Language (TMRQL). White Paper, Networked Planet Limited (2005)
7. Schwotzer, T.: Ein Peer-to-Peer Knowledge Management System basierend auf Topic Maps zur Unterstützung von Wissensflüssen. Doctorate Thesis, Berlin (2006), http://opus.kobv.de/tuberlin/volltexte/2006/1301/
8. Larman, C.: Applying UML and Patterns: An Introduction to Object-Oriented Analysis and Design and Iterative Development, 3rd edn. Prentice-Hall, Englewood Cliffs (2004)
9. Gamma, E., Helm, R., Johnsaon, R., Vlissides, J.: Design Patterns: Elements of Reusable Object-Oriented Software. Addison-Wesley, Reading (1995)
10. Garshol, L.M.: Towards a Methodology for Developing Topic Map Ontologies. In: Maicher, L., Sigel, A., Garshol, L.M. (eds.) TMRA 2006. LNCS (LNAI), vol. 4438, pp. 20–31. Springer, Heidelberg (2007),
    http://dx.doi.org/10.1007/978-3-540-71945-8_3
11. Ueberall, M., Drobnik, O.: Collaborative Software Development and Topic Maps. In: Maicher, L., Park, J. (eds.) TMRA 2005. LNCS (LNAI), vol. 3873, pp. 169–176. Springer, Heidelberg (2006), http://dx.doi.org/10.1007/11676904_15

---

[5] http://www-ist.massey.ac.nz/wop/

[6] http://www.key-project.org/

[7] Interested parties are referred to the *Git wiki*, which contains a plethora of related links at http://git.or.cz/gitwiki/GitLinks

12. Ueberall, M., Drobnik, O.: On Topic Map Templates and Traceability. In: Maicher, L., Sigel, A., Garshol, L.M. (eds.) TMRA 2006. LNCS (LNAI), vol. 4438, pp. 8–19. Springer, Heidelberg (2007), http://dx.doi.org/10.1007/978-3-540-71945-8_2
13. Vatant, B.: From Implicit Patterns to Explicit Templates: Next Step for Topic Maps Interoperability. In: Proc. XML (2002), http://www.idealliance.org/papers/xml02/dx_xml02/papers/05-03-06/05-03-06.html
14. Yin, M.: Organizational Paradigms. Draft, (last accessed May 25, 2007), http://wiki.osafoundation.org/bin/view/Journal/ClassificationPaperOutline2
15. Grønmo, G.O.: Representing ordered sets in Topic Maps. [sc34wg3] newsgroup posting (2003), http://www.isotopicmaps.org/pipermail/sc34wg3/2003-October/001881.html
16. Garshol, L.M.: The Linear Topic Map Notation: Definition and introduction, version 1.3 (rev. 1.23, 2006/06/17), http://www.ontopia.net/download/ltm.html
17. Noy, N.F., Musen, M.A.: Ontology Versioning in an Ontology Management Framework. IEEE Intelligent Systems, 6–13 (July/August 2004) http://doi.ieeecomputersociety.org/10.1109/MIS.2004.33
18. Niu, N., Easterbrook, S.: So, You Think You Know Others' Goals? A Repertory Grid Study. IEEE Software, 53–61 (March/April 2007), http://doi.ieeecomputersociety.org/10.1109/MS.2007.52
19. Schmitz-Esser, W., Sigel, A.: Introducing Terminology-based Ontologies. In: 9th International Conference of the International Society for Knowledge Organization (ISKO) (July 2006)
20. Wheeler, D.A.: Comments on Open Source Software / Free Software (OSS/FS) Software Configuration Management (SCM) Systems (2005), http://www.dwheeler.com/essays/scm.html
21. Wan, D.: CLARE: A computer-supported collaborative learning environment based on the thematic structure of scientific text. PhD thesis, Department of Information and Computer Sciences, University of Hawaii (1994)

Toward a Topic Maps Amanuensis

Jack Park

SRI International, Menlo Park, CA
jack.park@sri.com

Abstract. The CALO project at SRI International provides unique opportunities to explore the boundaries of knowledge representation and organization in a learning environment. A goal reported here is to develop methods for assistance in the preparation of documents through a topic map framework populated by combinations of machine learning and recorded social gestures. This work in progress continues the evolution of Tagomizer, our social bookmarking application, adding features necessary for annotations of websites beyond simple bookmark-like tagging, including the creation of new subjects in the topic map. We report on the coupling of Tagomizer with a Java wiki engine, and show how this new framework will serve as a platform for CALO's DocAssist application.

Keywords: CALO, DocAssist, topic map, social computing, machine learning, dashboard.

1 Background

SRI International has been developing the Cognitive Assistant that Learns and Organizes (CALO[1]) project. CALO began predominantly as a kind of *semantic desktop* application, one that provides many services to individual users. Services provided include applications such as mail, calendar, instant messaging, desktop indexing and clustering, task management, and others. CALO uses a particular *ontology* to organize the artifacts and activities of office productivity settings. In earlier work, we discussed a mismatch between formal ontologies and the day-to-day needs of office workers [10] where we first identified the opportunity to *federate* personal representations of artifacts of daily experience with those supplied in formal representations. That work provided a topic maps based foundation for the project we describe here. In the following chapters, we will describe a work in progress, in which we describe a way to use topic mapping to assist authors during the creative process by maintaining what we call *dashboard-like* ready access to information related to their work in progress.

As a service to authors of documents necessary for meetings and presentations, CALO provides a *DocAssist* application that we think of as a type of writer's amanuensis[2]. DocAssist is the subject of continual evolution, starting first with an

---

[1] CALO: http://calosystem.org/
[2] Amanuensis: a personal assistant.

L. Maicher and L.M. Garshol (Eds.): TMRA 2007, LNAI 4999, pp. 140–153, 2008.

agent capable of turning query strings into *de novo* PowerPoint presentations based on harvested existing presentations and user guidance. For example, an early trial of that software started with the query string "evolution" and searched a collection of presentations that included talks on Darwinian evolution and evolutionary programming. The program included four slides from talks on evolutionary programming and one on Darwinian evolution. We chose to add another slide that covered Lamarckian evolution to round out the new presentation; CALO was able to help us rapidly organize existing slides into a new presentation.

We followed the introduction of topic maps to CALO with our project *Tagomizer*[3] [7], an instance of a social bookmarking application. Tagomizer provided CALO with an online source of user-generated *learned* information resources. The project supported CALO's *learning in the wild* initiative where new information enters the system through external means. In this case, users *tag* web pages thought to be related to specific projects they are conducting while using CALO. Tagomizer performs two kinds of background maintenance of information resources available to the user: indexical functions in the subject-centric[4] organizational sense of topic maps, and domain modeling associated with ontologies and formal and informal models of domains. We describe modeling in more detail below.

DocAssist's new features will include web-based services in support of CALO users performing information-related tasks beyond those already available in Tagomizer. Tagomizer presently facilitates tasks that include the discovery and annotation of information resources related to specific projects; tags used while bookmarking websites are based on projects that CALO is helping users manage. Recent additions to Tagomizer facilitate user creation of new subjects of interest. These subjects may or may not be specific web pages already tagged; they may simply be new subjects that are important to some project or that are mentioned at some already-tagged web page. The new portal will facilitate annotation of existing subjects through linking, tagging, commenting, referred to here as processes of *annotation*. In cases where new subjects are discovered, the portal permits creating those new subjects, with subsequent annotations added by users and by search agents built into Tagomizer. As before, CALO, the desktop application, queries the portal through the web services provided and archives new information locally for DocAssist.

To motivate this discussion of DocAssist, we first sketch some background observations and thoughts that animate our design and implementation. We then discuss our implementation as it is evolving, and finally describe related work.

A CALO user's knowledge domains or universes of discourse are characterized as collections of heterogeneous information resources that cover diverse world views and complex subjects. We therefore take a particular interest in the methods and opportunities for formal modeling as applied in a topic maps framework. Here are the words of two scientists describing formal modeling in their domain [1]:

> "In explaining the functioning of cells we focus on *relations* among *objects*. The basis for this is the *principium individuationis* of

---

[3] Tagomizer: http://www.ai.sri.com/software/Tagomizer
[4] Subject-centric: topic maps aggressively organize information resources around individual subjects, merging representations where necessary.

Schopenhauer (1969) [2]: any object always and everywhere exists purely by virtue of another object. The principle of individuation constitutes the world of experience and phenomena: for *anything* to be different to anything else (or to have changed), either space or time or both have to be supposed. The objects we deal with in experiments are *material objects* subject to physico-chemical change. In modelling we deal with *mathematical objects* for which we can state propositions and derive theorems. The definition of a system as a set of interrelated objects follows naturally. We distinguish between a *natural system*, which is a collection of interacting material objects (physical, chemical or biological), and a *formal system* made up of mathematical objects, equations or propositions. Mathematical modelling is then the process by which we bring a natural and a formal system into congruence."

In our parlance, formal models include ontologies such as CALO's built-in office ontology, and mathematical structures, for example, category theoretic structures. We presently do not propose to create and manipulate category theoretic structures in DocAssist,[5] but we believe that an understanding of the modeling capabilities we must facilitate is well-informed by studying category theoretic descriptions, as for example [1]. To develop that thought, consider that, since topic maps are capable of providing organized indexes into collections of heterogeneous resources, they also support a type of modeling process that scaffolds representations of relationships among the objects, the subjects, of such universes. Category theory has been compared to set theory in this way: while set theory allows representations of members of a set, category theory permits representations of the *social life* of those members [3]. Describing the social life of category members means describing the way members interact with each other within the category. A similar capability in relational databases would manifest as relationships between tables, or between columns within tables.

The variant of topic mapping we practice is called *subject mapping*, based on the Topic Maps Reference Model (TMRM) [6] [4], where no specific XML or SGML dtd is specified; we are free to create *subject proxies* and the *subject properties* they contain as representations (proxies) of *subjects* found in mapped information resources. Subject proxies exhibit the form and behaviors of *frame-based systems* [5] of artificial intelligence (AI). If there are differences in the application of subject proxies and frames of well-documented AI systems, they might exist in subtle ways. The TMRM, for instance, perhaps uniquely *requires* that any subject property (slot) used in a proxy be, itself, declared as a proxy (subject) within the map; the underlying subject maps framework is one that is *self documenting* such that authors of any other subject map are free to examine the properties used in other maps in order to make determinations needed to *merge* subjects from one map to another. Self-documenting frameworks are not unique to the TMRM. We understand that ontologies, as for instance expressed in the OWL languages, are also self-documenting. Issues related to *merging*, however, are outside the scope of this paper, but there are differences in the ways in which subject maps and OWL ontologies deal with merging processes.

---

[5] Category theoretic structures: there is a modicum of evidence to suggest that subject map structures might be capable of being made applicable to category theoretic analytics.

[6] TMRM: The index into the document is found at http://www.isotopicmaps.org/tmrm/

The particular subject map implementation discussed here is our *TopicSpaces* open source (Java) project. We reported earlier [6], [7], and [8] on the evolution of TopicSpaces and its application, Tagomizer. Tagomizer is a social bookmarking interface supported by a subject map. The process of utilizing *social gestures* (known under the terms social computing, Web 2.0, collaborative filtering, tagging, and others) as a means of acquiring and mapping information resources related to new and existing subjects within the map is central to our work, and to the vision we are bringing to reality. A larger rubric with which to understand our work is that of *collective sensemaking,* where many individuals behave in collective and collaborative ways to share and grow their understandings of situations in which they are engaged.

We center this discussion on the collective sensemaking process known as *annotation.* We wish to *connect two dots*: topic mapping and annotation. To do so, we think first about the nature of annotation, and later look at the relational structures of topic mapping. David Price [16] lists three characteristics of annotations:

- Finer granularity than most content management systems (CMS) store resources (typically document level)
- Stored and managed separately from the content being annotated
- Created and viewed outside normal authoring processes

The first two points are significant in that they define topic mapping. Topic maps always are created outside the information resources; they are primarily indexical in the same sense as the "back of the book index." In that sense, topic maps exist as annotation systems. The last point is equally significant. Think in terms of temporal sequencing in the social life of information:

1. Information resources are created
2. Information resources become available for consumption
3. Information resources are indexed and mapped
4. Information resources are annotated

Steve Newcomb has described topic mapping as an *editorial process* [personal communication]; that is, those who craft topic maps are making editorial decisions as they perform the *ontological engineering* tasks of choosing ways to model subjects being included in the map. Annotations happen later in the life cycle. When a topic map is created, information resources are being annotated and indexed at the same time. At a later time, others will continue the annotation process, performing collaborative filtering. While collective sensemaking through annotations is an on-going process on the web today, we are interested here in those processes as they are performed on structures specific to topic maps. We now discuss collective sensemaking in the wild and tie that to the relational structures of subject maps. We later tie that discussion to our CALO project.

2   Architectures for Collective Sensemaking

The common vernacular (ontology) applied to the *social web* (see below) now includes terms such as *folksonomy, tagging, annotation,* and *Web 2.0.* For purposes of this discussion, we use the term tagging as a type of annotation, performed by user gestures typically with a web browser, where clicking a *bookmarklet*[7] while visiting one website transfers information about that resource to another website that provides a web service that accepts the information for later processing by the user. At Del.icio.us and at Tagomizer, a bookmarklet results in transfer of a webpage *title* and *url* back to the host where the user then adds comments and tags. A difference between Del.icio.us and Tagomizer lies in the *database* each system applies to the task at hand. We surmise that Del.icio.us applies a traditional relational database to the task of persisting tags, users, resources, and descriptions, among other things. Tagomizer saves the same objects, but in a different[8] way: Tagomizer's database is a subject map, with all entities organized as subjects together with a complex web of relationships that connects those subjects.

Recalling Schopenhauer's *any object always and everywhere exists purely by virtue of another object*, our discussion of *architectures* is necessarily slanted toward relational representation systems, While we are not endorsing Schopenhauer's, or any particular philosophical stance, we observe and follow those who seek holistic representation schemes, and that particular quotation animates our search for architectures that can maintain representations of subjects that are, themselves, described by attributes (properties) that exist along many dimensions. In the following paragraphs, we will describe the evolution of thinking that leads to our subject mapping approach in the DocAssist architecture.

The general way in which relational systems are described is through the simplified nodes-and-arcs diagrams sometimes known as *concept maps*. Figure 1 establishes a visual vocabulary for the following discussion.

Fig. 1. *Concept Map* graph

The graph represents objects a software entity creates and marshals: nodes and labeled arcs. A node, in this discussion, is a container of properties that describe the subject that node represents, including the name and perhaps a database identifier for the node, and perhaps numeric values locating the node in a view, and perhaps also lists of inbound and outbound arcs. If a node contains references to other nodes, indicating an arc makes the connection, that node reference might be a pair giving the node and arc label, for example, the following simple representation scheme:

```
Node: Joe
     Arc: {Sue, Married}
```

In any high-dimensional world of complex systems, we need to be able to represent more precisely the nature of the relationships. We need to be able to represent,

---

[7] Bookmarklet: a tiny Javascript *widget* (typically a button) saved at a browser that performs specific functions when selected by the user.

[8] Different way: we are forced to surmise here since we have no access to the software for examination of the tasks performed; we can only observe system behaviors on the web and make educated guesses about the software patterns applied.

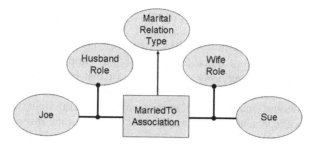

Fig. 2. Simple XML Topic Map

understand, and question more details than are represented in Figure 1. Topic maps entered the field, together with *conceptual graphs* [9], as graph structures intended to facilitate precise representation of concepts and their relationships. We speculated on the marriage of topic maps and conceptual graphs [10], but, for this discussion we will set aside the introduction of conceptual graphs and focus on the path from concept maps to subject maps. Figure 2 is the graph structure of an XML topic map according to the XTM specification.

We see that the topic map allows us to "say more things" and "ask more questions" about the situation modeled in the "Joe-Sue" marital relationship. There exists a class of questions we cannot ask in this particular graph, and that happens because the *association* itself is not a *subject* in the graph. For that, we have subject maps. Figure 3 illustrates a so-called *big-assert* structure of a subject map.

In the big-assert structure, we are able to make assertions about more aspects of the core relationship modeled between Joe and Sue. Joe and Sue each are *actors* in this relationship. The *casting* subject proxy is an optional subject that *casts* an actor (e.g., Joe) into a role and relationship. At the casting node linked to the Joe node in the graph, the particular subject is *the husband role Joe plays in a Joe-MarriedTo-Sue relationship*. The casting node can serve as an actor in other assertions (e.g., when the role started, when it ended, how it played out); the subject facilitates the notion of *seeing as* – we are able to model Joe *as* a husband. Other castings would place Joe in other roles.

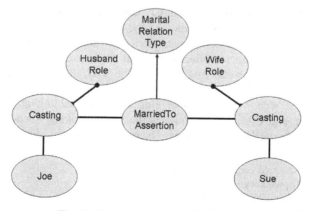

Fig. 3. *Big-Assert* graph of a Subject Map

While the big-assert model is an optional approach to TMRM subject maps, our description of that model illustrates an aspect of facilitating the high dimensional context space mentioned earlier. Consider the Joe-Sue example: it is not always sufficient to think in terms of a marital relationship as simply a relationship between two individuals; there are certainly other aspects of that relationship that will come into play; children, employment, finances, friends and other related people for starters. By providing subject-based anchors along several important dimensions within the primary representation, we facilitate extensions of the *story* encapsulated in each assertion along other dimensions.

3   Towards CALO's DocAssist Platform

The core application for this aspect of DocAssist is Tagomizer. Our migration from the Tagomizer project to DocAssist as we describe it here required that we search for a more flexible user interface than that provided by Tagomizer. The issue is this: Tagomizer supports only three dimensions along which a user can create and browse information: users, tags, and web pages. The new additions to Tagomizer call for presentation of subjects along perhaps infinitely many dimensions, or contexts. This calls for a more flexible user interface; we settled on the notion of joining TopicSpaces with a *wiki* framework. There are several candidate Java wiki frameworks from which to choose, including XWiki,[9] JSPWiki,[10] JAMWiki,[11] IkeWiki,[12] and several lesser-known projects. Each has its strengths; in the end, we are developing our platform on JAMWiki since it appears to provide the user experience of MediaWiki,[13] the software platform behind Wikipedia and many other websites. Figure 4 is Tagomizer now displaying in JAMWiki.

At this time in the Tagomizer evolution, the JAMWiki – Tagomizer marriage is progressing. We recently excised the original JAMWiki database from that code and rewrote parts of JAMWiki code to use TopicSpaces as the wiki's database. Doing so, we are asking TopicSpaces to behave as a CMS as well as a subject map. The system, thus far, is compact enough that we are able to run the entire platform from a one gigabyte USB memory stick. Doing so, we are able to explore issues of reliability of the platform under varying conditions.

A goal of the marriage is to seamlessly unite the presentation layer with the database, the knowledge layer. By removing the wiki database and writing code that turns wiki database queries into topic map queries, we are coupling users directly with the stored knowledge artifacts collected through Tagomizer and other means. This marriage occurs at the web-based server side; the wiki and TopicSpaces exist as a portal accessible by both users and CALO applications. Users perform social computing on the web, finding and tagging resources, adding new subjects and annotating those, and CALO, later, queries the server looking for new information with which to serve user's desktop needs. The combination of a web portal for CALO

[9] XWiki: http://www.xwiki.org/

[10] JSPWiki: http://jspwiki.org/

[11] JAMWiki: http://jamwiki.org/

[12] IkeWiki: http://ikewiki.salzburgresearch.at/

[13] MediaWiki: http://www.mediawiki.org/

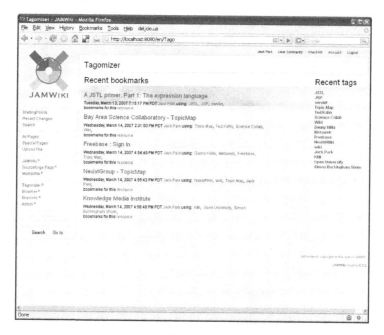

Fig. 4. Tagomizer in JAMWiki: CALO DocAssist taking shape

and individual CALO desktop installations facilitates collective sensemaking: many individuals contributing to a global knowledge base to be used by all.

4   Related Work

Numerous projects exist on the so-called *social web*. Kristina Lerman, at her website [12] says

> "The label 'social media' has been attached to a quickly growing number of Web sites whose content is primarily user driven."

While we will not attempt to enumerate all such sites here, many well-known sites are worth mentioning, including *Slashdot, Flickr, YouTube, Del.icio.us, digg*, and *Wikipedia*. Wikipedia is best known as a user-created collection of information resources about specific subjects. Slashdot and dig relate to social filtering of news items and events. Flickr exists for the collection of user-supplied images and tags, YouTube for movies and tags, and Del.icio.us for tagging everything else on the web, with no particular content of its own.

Another photo browsing and annotating service is CONFOTO[14] [13]. The project is an online service that moves closer to Semantic Web compatibility with Web 2.0-like interfaces. The site provides the ability to annotate photos, events, and people. The site is relatively new and undergoing changes; at this time, the site captures

---

[14] CONFOTO: http://www.confoto.org/

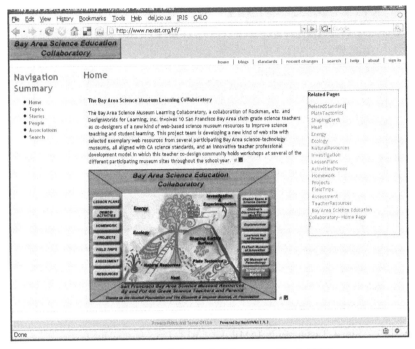

Fig. 5. Bay Area Science Collaboratory: NexistWiki

metadata about the artifacts represented, possibly including annotations, in RDF collections that look like the following snippet clipped from the site[15]:

```
<rdf:Description rdf:about=
    "http://www.confoto.org/media/user_78/2006/11/22/digns.jpg">
 <sk:localMediaPath rdf:datatype=
                "http://www.w3.org/2001/XMLSchema#string">
 user_78/2006/11/22/
 </sk:localMediaPath>
 <dc:subject>hans wurst test</dc:subject>
 <confoto:dateAdded>2006-11-22T11:38:43Z</confoto:dateAdded>
 <rdf:type rdf:resource="http://xmlns.com/foaf/0.1/Image"/>
 <dc:date rdf:datatype=
                "http://www.w3.org/2001/XMLSchema#dateTime">
  2006-11-22
 </dc:date>
</rdf:Description>
```

Our work with TopicSpaces follows our work with NexistWiki, our first implementation of a wiki-like architecture expressed as a topic map. The Bay Area Science Collaboratory is a project that provides online services to California science

---

[15] RDF snippet found at http://www.confoto.org/photos?r=http%3A%2F%2Fwww.confoto.org %2Fmedia%2Fuser_78%2F2006%2F11%2F22%2Fdigns.jpg&offset=   on 14 May, 2007 in the RDF Data tab

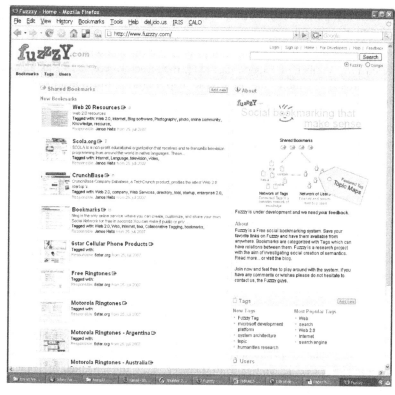

Fig. 6. Fuzzzy.com Social Bookmarking with Topic Maps

teachers situated in the Bay Area [14]. In one instance of the collaboratory (Figure 5), 6th grade science teachers enter links to their online course materials into the topic map; all course material is organized and related to specific California State curriculum requirements and to Bay Area science museum exhibits. Thus, teachers, students, and even parents are able to access course materials, which can include lecture notes, assignments, and activity plans, together with links to related science museum exhibits.

A topic maps-based social bookmarking portal, similar in many ways to our Tagomizer project is the portal fuzzzy.com[16] created by Roy Lachica for his master's thesis at University of Oslo, Figure 6, and well described in these proceedings [15]. Fuzzzy and TopicSpaces each share the distinction from del.icio.us and other social bookmarking websites as being applications of topic maps.

At the website, Roy says this about the project:

> "Fuzzzy is not only social but also semantic which means that the Tags used have more meaning. When bookmarks are assigned a meaning using a standard like the ISO 13250 Topic Map then people as well as

---

[16] Fuzzzy.com: http://www.fuzzzy.com/

other computer systems can make use of the embedded knowledge in a more meaningful way."

Relevant to our work is the ScholOnto project[17] of the Knowledge Media Institute. As reported in a number of papers, e.g. [11], the structure created in its annotation architecture greatly resembles that of a topic map. Figure 7 illustrates the nature of a ScholOnto link.

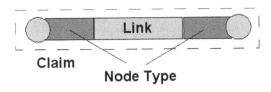

Fig. 7. ScholOnto Link

At each end of the diagram are resources serving as subjects about which a *claim* will be asserted through social gestures not unlike those of Del.icio.us or Tagomizer. The claim itself is the *link,* and the node types are similar to *roles* of a topic map. We believe that the ScholOnto project models its claims in the same way that a topic map models relationships between actors. ScholOnto assertions include *same as, supports, not same as,* and many others. Those assertions are organized in a simple ontology; ScholOnto thus applies a particular ontology to the social annotation process, and serves as a model for social annotation processes being implemented in TopicSpaces for our DocAssist project.

5 Discussion

The original Tagomizer project is being extended to provide capabilities and services necessary to create a web portal for CALO. We are modeling those new capabilities based on original ideas and those we are able to experience with related projects. Our interest in the related activities mentioned above lies in the nature of the information resources they marshal and the user experiences associated with the ways in which each of them operates. With the exception of the ScholOnto project, the contrast we see between each of these and our vision is the same difference we described earlier between Del.icio.us and Tagomizer. Our working hypothesis relates to the simple notion that aggressive maintenance of subject-centric organization of information resources can pay off in terms of improved user experience in creating, maintaining, and accessing the world's information resources. That is the central thesis behind this aspect of the DocAssist project.

Returning to the Tagomizer framework, the core difference between Del.icio.us and many of the existing tagging/annotation projects and Tagomizer is thought to be best summarized by the fact that the subject map database facilitates the federation of

---

[17] ScholOnto: http://kmi.open.ac.uk/projects/scholonto/

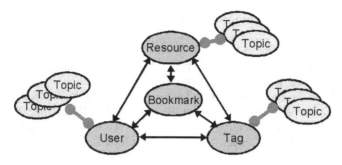

Fig. 8. Tagomizer and Related Subjects

additional information resources together with the tags, users, and resources those projects now marshal. Figure 8 returns to the concept map metaphor to illustrate the opportunity to federate other subjects together with those that are the focus of the website itself.

Figure 8, for simplicity, leaves out the spider web of associations, roles, castings, association types, and so forth. Each surrounding topic is either directly or indirectly related to some CALO project.

An interesting contrast is noted between the DAS approach of using individual clients to marshal resources found at various servers. The DocAssist subject map marshals information resources similarly, and forms a single source of federated information resources from which many CALO installations can access the resources needed.

ScholOnto contrasts with the Del.icio.us and Tagomizer tagging process since that project prescribes a particular tag set to use. Tags in ScholOnto are assertions made by a user; those tags assert particular relationships as defined in ScholOnto's ontology. The contrast lies in the lack of ontological prescription of tags at Del.icio.us, and non-CALO tagging with Tagomizer. CALO users, as mentioned earlier, use tags that are named according to specific projects. The contrast lies along a dimension that describes the amount of prior knowledge associated with particular tags. In this dimension in one direction are folksonomies, where one presumes that in the extreme, absolutely random naming events occur; in the other direction are ontologies of varying degrees of formality. This particular dimension remains the subject of continued research everywhere.

DocAssist as described here serves as a platform on which collaborative filtering, tagging and social gestures of various types can be applied to the task of improving the lot of CALO users. The same framework can be taken outside the CALO application and used in a variety of research and education projects. We believe that the federation of information resources in a subject-centric way will offer many benefits to a wide range of applications.

We see this contribution to CALO as opening the door to further research. We are building a platform for collective sensemaking, a field that includes many forms of discourse annotation. We expect to look at *dialog mapping* [17] as an appropriate next contribution to the subject mapping framework we are preparing for CALO.

CALO can then provide services to maintain annotated conversations as projects progress.

Acknowledgments. This work was supported by the Defense Advanced Research Projects Agency (DARPA) under Contract No. NBCHD030010. Any opinions, findings, and conclusions or recommendations expressed in this material are those of the authors and do not necessarily reflect the views of DARPA or the Department of Interior-National Business Center (DOI-NBC). This work has benefited greatly from conversations with Adam Cheyer, Ray Perrault, Joshua Levy, Eric Yeh, and Patrick Durusau, and two anonymous reviewers.

References

1. Wolkenhauer, O., Hofmeyr, J.-H.S.: An Abstract Cell Model that Describes the Self-organization of Cell Function in Living Systems. J. Theore. Biol. (2007), doi:10.1016/j.jtbi.2007.01.005
2. Schopenhauer, A.: The World as Will and Representation (first published in 1819). Dover Publications, Inc, New York (1969)
3. Fiadeiro, J.L.: Categories for Software Engineering. Springer, New York (2005)
4. Durusau, P., Newcomb, S., Barta, R.: Topic Maps Reference Model, 13250-5 (2007), http://www.isotopicmaps.org/TMRM/tmrm.1.33.pdf
5. Minsky, M.: A Framework for Representing Knowledge. In: Winston, P. (ed.) The Psychology of Computer Vision, McGraw-Hill, New York (1974), http://web.media.mit.edu/~minsky/papers/Frames/frames.html
6. Park, J.: Tagomizer: Subject Maps Meet Social Bookmarking. In: Maicher, L., Sigel, A., Garshol, L.M. (eds.) TMRA 2006. LNCS (LNAI), vol. 4438, Springer, Heidelberg (2007)
7. Park, J.: Promiscuous Semantic Federation: Semantic Desktops Meet Web 2.0. In: Cruz, I., Decker, S., Allemang, D., Preist, C., Schwabe, D., Mika, P., Uschold, M., Aroyo, L.M. (eds.) ISWC 2006. LNCS, vol. 4273, Springer, Heidelberg (2006)
8. Park, J.: Topic Mapping: A View of the Road Ahead. In: Maicher, L., Park, J. (eds.) TMRA 2005. LNCS (LNAI), vol. 3873, pp. 1–13. Springer, Heidelberg (2006)
9. Sowa, J.: Conceptual Structures: Information Processing in Mind and Machine. Addison-Wesley, Reading, MA (1984)
10. Jack, P., Cheyer, A.: Just for Me: Topic Maps and Ontologies. In: Maicher, L., Park, J. (eds.) TMRA 2005. LNCS (LNAI), vol. 3873, pp. 145–159. Springer, Heidelberg (2006)
11. Buckingham Shum, S.J., Uren, V., Li, G., Sereno, B., Mancini, C.: Modelling Naturalistic Argumentation in Research Literatures: Representation and Interaction Design Issues (Special Issue on Computational Models of Natural Argument, Eds: Reed, C., Grasso, F.). International Journal of Intelligent Systems 22(1), 17–47 (2007)
12. Lerman, K.: The Social Web, http://www.isi.edu/~lerman/projects/socialweb/
13. Nowack, B.: CONFOTO: Browsing and Annotating Conference Photos on the Semantic Web. J. Web Semantics 4(4), 263–266 (2006)
14. Kahn, T.: Science Museum Learning Collaboratories: Helping to Bridge the Gap Between Museums' Informal Learning Resources and Science Education in K-12 Schools. In: Museums and the Web 2007, San Francisco, California, April 11-14 (2007), http://www.archimuse.com/mw2007/papers/kahn/kahn.html

15. Dino, K., Lachica, R.: Towards holistic knowledge creation and interchange. Part II: Examples, theory and strategy (In these proceedings). In: Third International Conference on Topic Maps Research and Applications: TMRA (2007)

16. Price, D.: Annotation Management in XML Content Management Systems, A Case Study. In: XML 2002 Conference, Baltimore, MD, December 8-12 (2002),
    `http://www.idealliance.org/papers/xml02/dx_xml02/papers/`
    `06-03-01/06-03-01.html`

17. Conklin, J.: Dialogue Mapping: Building Shared Understanding of Wicked Problems. John Wiley & Sons, Ltd, Chichester (2005)

Cooperative Building of Multiple Points-of-View Topic Maps with Hypertopic

L'Hédi Zaher[1], Jean-Pierre Cahier[1], and Claude Guittard[2]

[1] Tech-CICO / ICD, Université de Technologie de Troyes.
12 rue Marie Curie, 10010 Troyes cedex, France
{zaher,cahier}@utt.fr
[2] BETA, Université Louis Pasteur.
4 rue Blaise Pascal, 67070 Strasbourg cedex, France
guittard@cournot.ustrasbg.fr

Abstract. This paper presents *socio-semantic web* applications based on the *Hypertopic* knowledge representation model. The Hypertopic [4] model is used to collaboratively construct topic maps with multiple user points-of-view within related communities. Three model building methods are presented herein: centralized, conflict-based, and hybrid. These methods are expressed using the socio-technical 'SeeMe' notation and method [7,8]. This paper will also illustrate how various co-building approaches can be used to construct Hypertopic maps in real communities.

Keywords: Design methodology, Knowledge engineering, Socio-semantic Web.

1 Introduction

Socio-semantic Web [3] applications make it possible for communities to co-produce and symbolically organize artifacts such as *topic maps*, *tag clouds* or *shared indexes*. A socio-semantic Web makes collective knowledge and activities more visible and more reflexive yet does not involve a high-level of automation due to the processing of formal ontologies by inferences. From the knowledge management point-of-view, the socio-semantic Web supports communities requiring a continuous and collective elicitation of information about the local semantic structures of business objects and their ongoing collective work. Projects of this nature involve the use of Web 2.0 collaborative applications requiring large-scale "architectures of participation" [10] such as the Open Directory Project, Del.icio.us, Flickr, or end-user cooperative classification and communication techniques based on shared metadata [9].

New approaches are needed for the co-building of large socio-semantic Web applications that employ focused and adaptable methods for each individual community. Furthermore, organizational rules must be affirmed depending on whether the semantic decision-making process is centralized or distributed, (*i.e.* whether decision-making is a top-down, bottom-up, or hybrid process, whether a moderator or facilitator is involved, etc.).

L. Maicher and L.M. Garshol (Eds.): TMRA 2007, LNAI 4999, pp. 154–159, 2008.
© Springer-Verlag Berlin Heidelberg 2008

Since 2002, the Tech-CICO lab has promoted the standardization of the Hypertopic concepts and schema[1] with several communities, such as 'Argos', 'Agoræ', 'Porphyry', 'Cassandre', using Tech-CICO Team developed[2] Hypertopic-based groupware tools to co-build socio-semantic Web applications. In these instances, many pragmatic methods have been used to "bootstrap" Hypertopic maps including the multiple view point functionality. The study group proposes to focus on three of these co-building methods used in Hypertopic applications built with Agoræ:

- a centralized co-construction method corresponding to a top-down approach with a semantic facilitator intervening at the bootstrap stage;
- a conflict-based co-construction method with a facilitator to make the conflicts more explicit at the bootstrap stage in small groups;
- a hybrid method associating top-down, centralized approaches with bottom-up decentralized approaches.

2 Socio-semantic Activity Guidlines

Community members require a set of socio-semantic activity guidelines which pertain to the launch and maintenance of topic map co-construction. These guidelines must articulate the specific models while combining three components: *a knowledge representation language* or 'metasemiotic' that community members can use to construct and discuss collective mappings. In order to provide a cartographic representation of items from multiple players' points of view, the Hypertopic model is needed (cf. §3); *an activity model* must be articulated with and upon the knowledge representation model. In particular, a model of roles must be supported by groupware tools to allow for the map to be cooperatively co-built; and *a method* precisely defining player practices including how and when they interact with socio-semantic activities.

The creation and the entire lifecycle of a socio-semantic Web cartography mobilizes players at both the socio-organizational and inter-personal levels. Methods of assisting these activities must, therefore consider a wide-range of socio-technical aspects. Thus, a special reference is made to recent studies on distributed collective practices and Computer-Supported Cooperative Work (CSCW) [1][9][12][15].

At the community level, the local semantic system is continuously and collectively constructed by the players as they perform their activities. In this *autopoietic* process, users are not only passive consumers of externally-designed semantic resources; they are users and creators of local semantic resources that they are able to manage by themselves using, for example, Web standards. Whenever it is necessary to explicitly define parts of the underlying semantic resources, the best, or in many cases, the only solution is therefore for semantic structures to be managed by the stakeholders involved in line with the 'participatory design' principle. This applies especially to

---

[1] See [18] and http://www.hypertopic.org
[2] See: http://sourceforge.net/projects/argos-viewpoint/; http://www.porphyry.org/;
  http://source-forge.net/projects/cassandre-qda/; http://sourceforge.net/projects/agora

situations where the underlying semantic resources which need to be elicited and updated are particularly voluminous, evolutionary, and even conflict-based.[1].

3 Hypertopic

Hypertopic models [4,17,18] were designed to represent knowledge about a collection of *items* such as *products* (in the case of a virtual marketplace), *R&D projects* (catalogue), *persons* (map of skills), *books* (in a library), etc. Hypertopic provide the formal framework in which shared meaning is explicitly described by means of a topic map with multiple points-of-view . Given the need for re-usability, Hypertopic include several basic constructs which allow end-users to understand a map. The Point-of-View (PoV) is a descriptor which serves to contextualize an item to illustrate the relevant actors and/or the main aspects involved in its analysis.

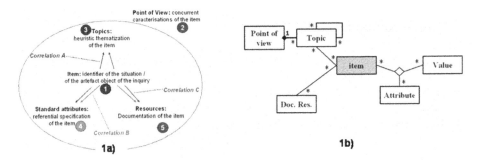

Fig. 1. The Hypertopic Model: **1a)** Semiotic View on the Item; **1b)** UML notation

Hypertopic models can be considered as a particular template of Topic Maps [11][14] and are more constrained than the TMO standard [14]. Although partially inspired by the TMO standard for topics, associations, internal/external occurrences, and scope, the basic constructs of Hypertopic (Figure 1b) add methodological constraints, to be easily understood by the community of users and to give to many end-users the ability to edit maps without prior training. For example, as a PoV is a high-level element, users are restricted from creating a lower-levelled scope. The PoV, therefore, makes use of the user position or homogeneous Knowledge Management collective choices within the organization. To fix the usage by simplifying the knowledge representation makes it easier to deploy the socio-semantic activity. Note that Hypertopic Points-of-View can be either have multiple opinions with conflict-based plurality or multiple dimensions with analysis within a particular user's PoV. This is generally for when more precise methods are needed (see §3) specifying which kind of PoV is involved in any given method and how & when users should build them.

4 Three Socio-semantic 'Bootstrap' Methods

We focus our methodological effort towards the more difficult "Bootstrap" phase of socio-semantic activity. This phase furnishes a set of points-of-view on a particular

Table 1. Three applications with quantified data

Socio-semantic Web applications : *Reference papers for more details :*		YePOSS **[2]**	DKN **[16]**	CartoDD **[3]**
Map	Points of view	6	3	7
	Topics	264	761	~500
	Items	169	75	~50
	Documentary resources	-	75	~200
Community	Contributors	7	3	~20
	Total users	-	12	-

item. However, controversies can arise during the co-designing phase especially during the early 'bootstrap' stage. The challenge here would be to develop in parallel the shared community culture and the basic set of terms, standards and rules resulting from the often implicit knowledge of members. In the following diagrams, the "SeeMe" modelling method [7,8] is used to make the co-building methods more explicit with the sequencing of numbers showing the scheduling of actions.

The centralized co-construction method (Figure 2) involves the intervention of a semantic facilitator during the bootstrap phase. This method is repeatedly tested in several cases using the Knowledge-Based MarketPlace co-construction model [1] for example the Yeposs (Yellow-pages for Open Source Software) experiment [2].

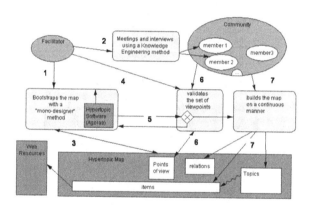

Fig. 2. Centralized Co-Building Method

The conflict-based co-construction method of Figure 3 below does not require a facilitator. In the DKN SeqXAM study case [16], groups attempting to share knowledge were liable to experience cognitive conflicts [13] due to the existence of various points-of-view, semantic heterogeneity, or interpretative disagreement (see demonstration at http://tech-web-n2.utt.fr/dkn/).

Finally, the hybrid method (Figure 4) blends a centralized design with a wiki-like, bottom-up approach containing tag-clouds coming from multiple users. This method

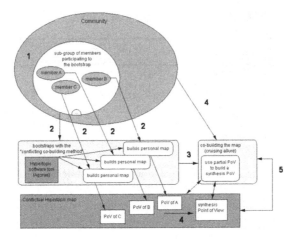

Fig. 3. Conflict-Based Co-Building Method

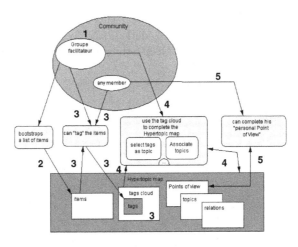

Fig. 4. Hybrid Co-Building Method

is presently used for the co-building by stake-holders of a map in the field of sustainable development [3] (see demonstration at http://tech-web-n2.utt.fr/dd/).

5 Conclusion

This paper introduces methods for the collaborative construction of socio-semantic Web applications based on Hypertopic. Future works entail the integration of more compete tools for co-operation inspired by the CSCW. Of further interest is the use of topic maps in open information retrieval [3] practices.

References

1. Cahier, J.-P., Zacklad, M.: Towards a Knowledge-Based Marketplace Model (KBM) for Cooperation Between Agents. In: Proc. of COOP 2002 conference. IOS Press, Amsterdam (2002)
2. Cahier, J.-P., Zaher, L'H, Leboeuf, J.-Ph., Pétard, X., Guittard, C.: Experimentation of a Socially Constructed Topic Map by the OSS Community. In: Proc. of the IJCAI 2005 workshop on Knowledge Management and Ontology Management (KMOM), Edimbourg (2005)
3. Cahier, J.-P., Zaher, L'H., Zacklad, M.: Information Seeking in a Socio-Semantic Web Application. In: Buckingham Shum, S., Lind, M., Weigand, H. (eds.) Proc. of the 2nd International Conference on the Pragmatic Web. ACM International Conference Proceeding Series, vol. 280, pp. 91–95. ACM, New York (2007)
4. Caussanel, J., Cahier, J.-P., Zacklad, M., Charlet, J.: Les Topic Maps sont-ils un bon candidat pour l'ingénierie du Web Sémantique? In: Conférence Ingénierie des Connaissances IC 2002, Rouen Mai (2002)
5. Dewey, J.: Logic, the theory of the Inquiry. In: Boydston, J.A. (ed.) The Collected Works of John Dewey, (pp. 1882-1953, Carbondale and Edwardsville: Southern Illinois University Press, pages 1969-1991), The Later Works - vol.12: (LW.12) 550 (1938)
6. Ehn, P., Badham, R.: Participatory design and the collective designer. In: Binder, T., Gregory, J., Wagner, I. (eds.) Proc. of the Participatory Design Conference, pp. 1–10 (2002)
7. Herrmann, Th., Loser, K.-U.: Vagueness in models of socio-technical systems. Behaviour & Information Technlogy 18(5), 313–323 (1999)
8. Herrmann, T., Kunau, G., Loser, K.-U.: Socio-Technical Self-escriptions as a Means for Appropriation. In: Workshop Supporting Appropriation Work: Approaches for the reflective user, ECSCW 2005, Paris, France (2005)
9. Mathes, A.: Folksonomies, Cooperative classification and communication through shared metadata. Univ. of Illinois Urbana-Champaign (2004)
10. O'Reilly, T.: What Is Web 2.0: Design Patterns and Business Models for the Next Generation of Software (2005), http://www.oreillynet.com/pub/a/oreilly/tim/news/2005/09/30/what-is-web-20.html
11. Park, J., Hunting, S.: XML Topic Maps: Creating and Using Topic Maps for the Web. Addison-Wesley, Reading (2002)
12. Schmidt, K., Wagner, I.: Ordering systems: Coordinative practices and artifacts in architectural design and planning. In: Proc. of the International ACM SIGGROUP Conference on Supporting Group Work (GROUP 2003), pp. 274–283. ACM, New York (2003)
13. Simmel, G.: Conflict & the Web of Group Affiliations. Free Press, New York (1955)
14. XML Topic Maps (XTM) 1.0, TopicMaps.Org Specification (2001)
15. Zacklad, M.: Communities of Action: a Cognitive and Social Approach to the Design of CSCW Systems. In: Proc. of the International ACM SIGGROUP Conference on Supporting Group Work (GROUP 2003), pp. 190–197. ACM, New York (2003)
16. Zaher, L'H., Cahier, J.-P., Turner, W.A., Zacklad, M.: A conflictual co-building method with Agoræ. In: Proc. of Workshop on Knowledge Sharing in Organizations, COOP 2006, Carry le Rouet, France (2006)
17. Zaher, L'H., Cahier, J.-P., Zacklad, M.: Information Retrieval and E-Service: Towards Open Information Retrieval. In: Proc. of International Conference on Service Systems and Service Management IC SSSM 2006, pp. 41–46. IEEE, Los Alamitos (2006)
18. Zhou, Ch., Lejeune, Ch., Bénel, A.: Towards a standard protocol for community-driven organizations of knowledge. In: Proc of the 13th ISPE International Conference on Concurrent Engineering (ISPE CE 2006), pp. 338–349. IOS Press, Amsterdam (2006)

Metadata Creation in Socio-semantic Tagging Systems: Towards Holistic Knowledge Creation and Interchange

Roy Lachica[1] and Dino Karabeg[2]

[1] Bouvet ASA, Pb 4430 Nydalen, 0403 Oslo, Norway
roy.lachica@bouvet.no
[2] The Department of Informatics, University of Oslo
dino@ifi.uio.no

Abstract. Fuzzzy.com, a social bookmarking website has been developed to study collaborative creation of semantics. In a shared online space, users of Fuzzzy continuously create metadata bottom-up by categorizing (tagging) favourite hyperlinks (bookmarks). The semantic network of tags created by users evolves into a people's fuzzy common ontology ("folktology"). We discuss several social and cognitive aspects of Topic Maps technology and scalability by analyzing the use of the system. We further argue that holistic knowledge creation and interchange is highly needed. Our results from Fuzzzy suggest that this can be realized by connecting distributed knowledge centric communities of dedicated users within specific domains.

1 Introduction

Studies have shown there is an ongoing reluctance among both users and institutions to create metadata [1]. The reluctance towards metadata creation causes the Web to sink into a morass of information overload and become a source of frustration and for many users.

There is also the need for existing metadata to be updated. Manual creation and updating is costly. Automatic processing often leads to poor quality because it is still suboptimal compared to human reasoning [2].

In dynamic and evolving knowledge centric communities knowledge structures must be able to evolve and adapt. Semantic Web research has revealed that one of the most challenging tasks has proven to be the development and maintenance of ontologies. Several languages now exist for computer mediated ontologies, but the creation and managing of these ontologies is time-consuming, difficult and often requires the involvement of both domain experts and ontology engineers [3], [4], [5], [6], [7]. In recent years we have seen diverse research targeting ontology creation and management with approaches ranging from automatic inferencing, to ontology engineering methodologies to collaborative environments for achieving consensus on ontologies [8]. Among the most widely researched approach to ontology creation is the Self-annotating Web paradigm [9] with the principle idea of using the available data of the Web to automatically create semantics.

Our approach to the problem of ontology evolution is the pragmatic approach of the Socio-semantic Web (S2W), which relies on flexible and evolving description languages for semantic browsing [10]. S2W emphasizes the importance of humanly created loose semantics as a means to fulfil the vision of the Semantic Web. Instead

L. Maicher and L.M. Garshol (Eds.): TMRA 2007, LNAI 4999, pp. 160–171, 2008.
© Springer-Verlag Berlin Heidelberg 2008

of relying entirely on formal ontologies and automated inferencing, humans, aided by socio-semantic systems, are collaboratively building semantics [11].

Folksonomies [12] have become widely popular in recent years because of their ease of use. Folksonomies and ontologies can be placed at the two opposite ends of a categorisation spectrum. The process commonly known as "tagging" has proven to be effective for creation of metadata. However, the quality of metadata created by folksonomy tagging is poor [13], [14], [15]. Also, current folksonomies used by popular sites such as Del.icio.us and flickr.com do not allow for sharing tags between applications [16]. Fuzzzy.com, described in this article, is the result of a semantic adaptation of the folksonomy which we label a 'folktology'.

Our contributions are two-fold. On the one side, we draw insights from the experience with Fuzzzy to discuss the feasibility of folktologies. On the other side, we develop the folktology approach further and show how this approach can be used as a basis for holistic knowledge creation and interchange.

The rest of this paper is organized as follows. Section 2 describes Fuzzzy.com and its folktology. In section 3 we evaluate the ontology-near categorization method of the Fuzzzy folktology by comparing it against folksonomies. We then go on to discuss the unsolved issues of the Fuzzzy folktology, the main of which is the persisting reluctance against metadata creation. In the fourth section we lay out a proposed strategy to tackle these issues.

2 Fuzzzy.com

The main concepts of the Fuzzzy system are bookmarks, tags and users. Bookmarks are created and tagged by users. By the end of October 2007 Fuzzzy had 221

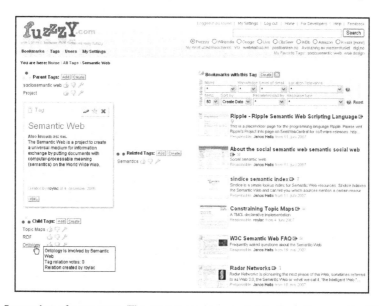

Fig. 1. Screenshot of a tag page. The current tag is presented in the yellow panel with related tags above, beneath and to the right. The right side of the screen shows a list of bookmarks that are tagged with the current tag.

registered users. Tags are contained in a semantic network created collaboratively by end-users. The mesh of users and the semantic network of tags becomes the folktology (folk + ontology).

Bookmarks can be recommended and saved as favourites. Tag to tag relations and tag-bookmark relations can be voted up or down. Users can find bookmarks by searching, filtering, by browsing the tag-space or by navigating the tripartite bookmark-tag-user page setup.

2.1 System Overview

Fuzzzy is an Asp.Net Ajax web application using the Networked Planet TMCore Topic Maps engine. Fuzzzy.com can be used with any modern web browser having JavaScript enabled. It has an experimental web service interface that enables it to act as a tag server and also to connect to other tag servers allowing for distributed global

Fig. 2. The folktology consist of any words that users choose to tag bookmarks with. All folktology tags and relations are instances of classes that are part of the tag ontology which is constructed from Topic Maps elements.

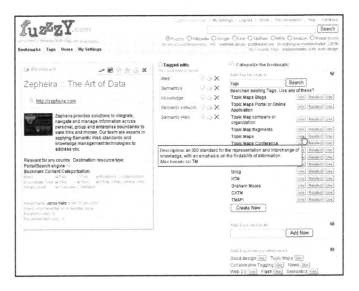

Fig. 3. Screenshot of a bookmark page. The left side shows a bookmark. The middle column shows a list of tags used for the bookmark. The right column shows part of the tagging functionality. The user has entered part of a tag name in the "Add tag by search" text–field. Tags with names or synonyms starting with the entered text are listed.

tagging across applications. Fuzzzy is a hybrid Topic Maps solution where the database contains both a topic map and other Fuzzzy specific intermediate data. To simplify the act of creating semantics, a minimalist core tag ontology scheme has been designed. The tag ontology consist of the 'Tag' topic type, topic types for specifying either vertical parent, vertical child or horizontal tag relations and 22 predefined association types each with a role player pair.

3 Evaluation of Semantic Creation in Fuzzzy

In folksonomy based systems, tag-to-tag relations are inferred by the tags different users have applied to the same resources. In the Fuzzzy folktology on the other hand tag-to-tag relations are explicitly added by users. We now discuss several issues related to this significant difference.

3.1 Comparison of Folktologies and Folksonomies

A preliminary study [17] showed an increase from an average of 32 % meaningful relations on 3 typical folksonomy sites to 97 % on Fuzzzy.com where a folktology is used. To evaluate tag relations we have devised a qualitative semantic relevance judgement method similar to that of Miller and Charles' contextual correlates of semantic similarity experiment [18]. In Miller and Charles experiment, semantic distance was measured by individuals rating contextual similarity for pairs of nouns. In our study the evaluation is done by articulating a relation between two given tags. If we can not clearly picture a relationship and specifically describe it verbally we assume the relationship is either faulty or the semantic distance is to long and should not be presented as a direct related tag. The criteria for a meaningful tag relation are: 1. the related tag must exhibit an appropriate level of specificity given a general purpose context. 2. The related tag must be unambiguous and readable. 3. The reader should be able to describe a relation verbally in normal spoken language within 10 seconds. 4. The relation must be intuitively grasped by a reader having basic understanding of the two tags.

While tags on Fuzzzy.com have fewer related tags in most cases, the semantic quality of tag relations is significantly higher.

The high number of non meaningful tag relations in folksonomies can be explained by the way in which users create ambiguous tags and use the pairing of tags in search.

Table 1. We have collected a total of 1632 tag pairs, from 4 different social bookmarking sites. Duplicate related tags, synonyms or plural/singular tags have been removed. Only tags with 2 or more relations are used and only the top 10 relations are included.

Bookmarking service	Tagging system	Tag relations evaluated	Meaningful tag relations
del.icio.us	Folksonomy	465	30%
bibSonomy.org	Folksonomy	462	30%
blogmarks.net	Folksonomy	470	37%
fuzzzy.com	Folktology	235	97%

In the evolving Fuzzzy folktology users are encouraged to collaborate to add more appropriate tags and to vote for the best tags for a resource. Folksonomies have no provisions for narrowing terms and the system have a tendency to be dominated by a few frequently used tags. A study by Tonkin [19] showed the unbalanced tag distribution of folksonomies.

Folksonomies have the potential to exacerbate the problems associated with the fuzziness of linguistic and cognitive boundaries [13]. The folktology reduces this problem by introducing semantics, synonym control and a collaborative environment where users can garden the shared tags. The Fuzzzy folktology also allow tag merging. Several tag pairs have been merged on Fuzzzy.com. One of these tag pairs was 'Web 2.0' and 'web2'. In folksonomies, categories can not be renamed. The tag is a free text property annotation. In the Fuzzzy folktology the tag is a standalone subject proxy that can be renamed. Synonyms can be added and the tag itself can be merged. In the merge process from the case above, 'web2' was added to the primary tag's list of synonyms. Tag merging in Fuzzzy is currently only available to administrators because the role and trust management module in not fully developed. Tag merging is a critical operation that may affect numerous users as tags are shared throughout the application by all users.

Folksonomies are suitable for serendipitous browsing and discovery of information as they reveal the digital equivalence of "desire lines" [20]. In folksonomies the poor semantics of tags often result in ambiguous but popular information as replacements for relevant information. A resource is indirectly recommended by the number of users who save the item. Users often experience the resulting ambiguity as a "nice to have" feature rather than a limitation.

3.2 Fuzzzy.com Unsolved Issues

Information Overload. With Topic Maps scopes [21], different contextual view-points can be expressed. Since contextual scoping allows for restricting the amount of information which is simultaneously visible, dividing information into specific views reduces the information overload. In the widest sense of an open social setting such as Flickr, Del.icio.us or Fuzzzy there is no defined domain of discourse and no precise division of users. In these environments there is no defined target group. Also, no two persons share identical world views [22]. This leads to a problem of defining scopes because scopes can not be adjusted to the needs the users. A World War II expert will have different needs for scoping on time than a palaeontologist. Most users might be comfortable with scoping content into English, French, German and Spanish while a language expert might need more fine grained language dialect scopes which is redundant to the average user.

As with scopes, Topic Maps types can not be decided collectively in the folktology making it hard to define a navigational structure. The folktology of Fuzzzy consists mostly of categories or types itself represented by tags. All of these tags are candidates for Topic Maps types, but any top-down structuring by choosing a set of main types in a folktology will have deep impacts on further use.

When presenting unscoped and untyped information in open folktologies, the danger of information overload becomes apparent as there is little structure and different levels of discourses are blended. Without scopes the user runs the risk of being overwhelmed by vocabularies that he or she is not comfortable with in addition to vocabularies that may already be contradicting.

Fuzziness of Socio-semantic Information. In our socio-semantic application the consensus view is constantly evolving. As culture and language evolve so does the folktology, and therefore the potential for overlapping, faulty or imprecise information is large. Some information will increase noise, not only because users use different vocabularies, but also because users make both semantic and syntactic mistakes. Casual users can not be expected to add precise and accurate information.

Low Degree of Participation. Interviews suggest that users did not perceive semantic metadata creation through the creation of tag relations as supportive of his or her personal goals. This is inline with Preece [23], which states that online communities must have a clearly stated goal.

Goal setting helps to gather users who are more in tune with each other and will better function as a whole. In Fuzzzy there is no community with a common goal to gather around. Creating semantics does neither support collective or personal goals and the core bookmarking purpose is not supported in the most optimal way because the system is designed for semantic creation and collaboration. Only a small minority of users created relations between tags. Users did not have the motivation to learn how to do it and they did not see any benefits from doing it. Users already have a mental representation of the world and have no need to externalize this view by entering their world view into the system.

Users of a bookmarking system require fast submission of bookmarks and fast access to them. Users often prefer to save bookmarks instantaneously without going through the process of adding metadata.

We have observed several cases where users do not agree with tag-relations that have been created by others but they take no action to correct it by voting or other means. Users are detached from the folktology and have no interest in seeing to that the folktology evolves into something that supports their views.

Irrelevance. Our usage logs showed that few bookmarks were voted as important. Bookmarks voted as important and displayed on the first page of Fuzzzy were seldom used. Users are most often only interested in bookmarks that are relevant to their specific context and situation or their personal interests and beliefs. The personal views of users on Fuzzzy.com are seldom shared and lists of important links, users and tags have no context and never become interesting for the reader. This increases the amount of irrelevant information, which leads to noise.

4 Towards Holistic Knowledge Creation and Interchange

We now take a wider look at the problems of informing on the Web and propose a new information infrastructure based on the results from Fuzzzy. We argue that the proposed infrastructure can solve problems uncovered with the Fuzzzy socio-semantic folktology and will bring us closer to realizing the vision of holistic knowledge creation and interchange.

4.1 The Need for Better Informing

The Internet consists of vast amounts of information that can lead to insight and knowledge needed for human and cultural as well as scientific development. Unfortunately, with the current information infrastructures, humans are often unable to locate the relevant resources. Information is mostly available but because it is hard to retrieve, people are willing to sacrifice information quality for accessibility [24].

The anarchistic architecture of the Web has enabled an explosive adoption which we have benefited greatly from. This architecture has now outplayed its role. The now prominent problem of information overload suggests that we need a "top of the mountain"-view of information, an infrastructure that can present the 'big picture' and highlight what is most relevant and credible [25].

With Web 2.0 applications and the growing blogosphere the Web is becoming more participatory. As more and more people can publish information, the Internet becomes more fragmented, with countless islands of discourse. Only 15% of web pages include links to opposing viewpoints [26].

As the society develops, diversity and complexity are added to the ever growing sea of information and new specialized research areas emerge. This suggests a need to shift from the present-day reductionist focus to a concurrent unified and holistic view. With the problems of growing complexity, information overload and the anarchistic Web we argue that there is a need for collaborative and democratic systems that can provide relevant and important information with a clear and correct view of the whole [25], tools that enable participants to discover truths and induce new knowledge.

4.2 From Folksonomies to Organic Ontologies

We hypothesize that the reluctance to create metadata can be diminished by adapting the folktology approach to become tools for semantic metadata creation in knowledge-centric communities.

In order to develop the folktology approach we first provide an analysis of the issues uncovered in the Fuzzzy folktology. Without a defined domain of discourse any information can be entered into the system increasing irrelevant information and the amount of noise for users. Few users share a sense responsibility to the metadata or the information content, and therefore users will not take the time to garden the information space. Irrelevant and large amounts of information along with noise cause a reduced user experience which results in few users adopting the system. This bootstrapping problem is also a result of design that is not aligned with the needs of the users. In Fuzzzy the creation of semantics is shadowed by the primary purpose of adding and retrieving bookmarks. Without no purpose and immediate benefits directly associated with the creation of semantics, few users will spend time on this activity. Without a community there is no discourse and therefore no domain of discourse (ontology) can develop.

By providing tools that can help to evolve knowledge within knowledge centric communities, both the community and the domain of discourse will be able to flourish. In these socio-semantic settings the folktology changes from a common universal fuzzy ontology, to specialized organic ontologies. When oriented towards smaller well defined communities of interest, the user interface can be designed to meet users' needs, reducing cognitive load and increasing information relevance. With a smaller community sharing a common goal, purposeful scopes and types can be decided, adding structure and reducing information overload by contextualization. Fuzziness is reduced when users share a relatively consistent vocabulary. Noise is reduced because dedicated users are more willing to do gardening work on the semantics.

Ontology creation and evolution is a time-consuming process which requires comprehension, analysis, synthesis and evaluation [28]. Dedicated users are needed for such activities. In a mature community, users will have the incentives to add valuable metadata. Motivation is increased when users feel unique and contributing [29]. Smaller communities have obvious advantage in this regard.

4.3 Holoscopia – Holistic Knowledge Trough Distributed Online Communities

The Holoscopia platform is conceived as a future system that can help world-wide deployment of ontologies which is seen to be essential for the growth of the Semantic Web [30].

The infrastructure of Holoscopia consists of interconnected Polyscopic Knowledge Bases (PKB). Each PKB is an autonomous knowledge base for polyscopic structuring [27] of information. The PKB is used by a community to evolve ideas, to develop a consensus about the knowledge within a domain, to decide what information is important, what are the key-points or wisdom that needs to be communicated to the larger community outside, and what actions this insight should lead to. The PKB is a democratic knowledge creation environment. It has an evolving organic ontology and acts as a portal into the collective community knowledge. It also let users browse the interconnected web of Holoscopia. Users of a PKB can import knowledge in the form of Topic Maps fragments from other PKB's.

Holoscopia provide holism by connecting and aggregating knowledge from the diverse communities and letting users explore intricate relations that can infer new knowledge. The PKB can be seen as a combination of a Wiki, bookmarking system, issue tracking, forum and hypothesis testing, decision support and concept mapping tool. The functionality supports mapping, organizing and discovering the complexity of reality. The resulting wisdom of a PKB can serve as guiding principles for motivating direction change or encourage formal scientific studies. In a utilitarian sense, the collective wisdom facilitated by Holoscopia can help us to foresee consequences and thus leveraging the long-term common good.

Use Case Scenario for a Polyscopic Knowledge Base. A typical use of the PKB could be within the Topic Maps community. Members might use the system to discuss what areas need more research and why, what aspects of Topic Maps needs to be tested in real world applications etc. The system helps users to systematically build a line of arguments with supporting resources or previous statements that have been agreed upon. Users build a consensus map of the domain through the organic ontology. The resulting knowledge base can be connected to other domains such as the domains of Natural Language Processing or the Semantic Web. Synergy can be achieved when users view the updated essence of neighbouring domains and get new ideas or find information or knowledgeable members that can help to solve problems in their own domain.

Other typical uses for the system are within global social issues such as poverty, globalization, climate change etc. where the problem is complex, broad and fuzzy. With Holoscopia these issues can be investigated by the public.

Social Layer. Similar to the HyperTopic model [31], the organization of users and their activities must be facilitated by the system. Coordination of users and their tasks becomes crucial as knowledge is collectively and continuously constructed. For a collective knowledge corpus to thrive there must be mechanisms for correcting erroneous knowledge and for enhancing and collaboratively evolving the existing knowledge. Morville [32] points out there is a fine line between wisdom of the crowds and ignorance of the mob. This brings us to the question of whether an elite or the crowd is most fit to make collective decisions on behalf of the community.

Based on principles of enlightened democracy [33] we propose a democratic election model for online communities. Users who carry out voluntary work receive credits. Users vote on others which they believe are important, knowledgeable or in any way beneficial to the community. The user mass evolves into a community of members with different levels of trust and influentiality. The more trust a user has earned, the more privileges will be granted to him or her. All users have the right to create or import new information. Users with medium trust level can update information added by others. Users on higher levels have permissions to merge and delete. Users at the top level have policing rights and can suspend low level users.

Intermediate Knowledge Layer. In our application of collective knowledge synthesis, conceptualizations may need to be modified as new knowledge is gained and the world views of members change. Therefore it is not only the content contained in the PKB that evolves but also the knowledge structure. When an ontology reaches a certain size and complexity, the task of removing outdated parts and adapting valid parts becomes huge [34]. We introduce the notion of organic ontology as a metaphor to describe the user-created evolving ontology.

In the organic ontology, topics are nurtured through their use and trough gardening work. Topics that are not cared for will die. Topic relations grow stronger as they are used. Relations that people find inappropriate will be gardened out by voting. As in Darwinian evolution, the fittest topics and relations survive and gain visibility, while the others that are unused, or with negative votes go extinct. In the current version of Fuzzzy, unpopular tag relations are ranked lower in lists and when reaching a lower threshold they will not be visible in default views.

The symbiosis between the community members and the living ontology lowers the effort that is required of the members and also decrease information overload.

Semantic Layer. The semantic layer consists of a Topic Maps engine and holds one topic map instance. The topic map reflects the current consensus knowledge, based on member input, filtered and processed by the intermediate layer.

The semantic layer contains mechanisms for machine readable knowledge interchange and interoperability. To allow for both knowledge interchange and evolution, a core ontology is needed. Similar to figure 2 we propose a user created ontology on top of a master ontology using he Topic Maps constraints language.

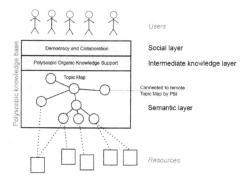

Fig. 4. The three layered architecture of a PKB

5 Conclusion

We have seen that collaborative approaches to simple ontology-near evolution are possible in open online environments without dedicated resources such as ontology engineers. The Socio-semantic Web approach can be used to create semantics organically from a relatively small core without a single authority.

Demonstrated by Fuzzzy.com, the Topic Maps based folktology enables sharing of tags between applications across the Web.

Our study has shown that folktology tagging increases the semantic quality of categorization compared to folksonomy tagging. However, folktology tagging is more time-consuming and few users are willing to create semantics in a general social bookmarking application.

The problems of large scale open and common socio-semantic systems can be summarized by; information overload, fuzziness, low participation and a large amount of irrelevant information. The ongoing development of the Holoscopia platform aims to solve these issues. We argue that, by providing the appropriate tools and infrastructure to accommodate knowledge centric communities, semantic metadata can grow as a by-product of dialog and discovery on a global scale.

References

1. Casey, J., Proven, J., Dripps, D.: Geronimo's Cadillac: Lessons for Learning Object Repositories 7 (2005)
2. Marshall, C.C., Shipman, F.M.: Which semantic web? In: Proceedings of the Fourteenth ACM Conference on Hypertext and Hypermedia, Nottingham, UK, August 26-30 (2003)
3. Dicheva, D., Dichev, C.: Authoring Educational Topic Maps: Can We Make It Easier? In: ICALT, pp. 216–218 (2005)
4. Domingue, J., Motta, E., Shum, S.B., Vargas-Vera, M., Kalfoglou, Y., Farnes, N.: Supporting Ontology Driven Document Enrichment Within Communities of Practice. In: International Conference on Knowledge Capture, Victoria, British Columbia, Canada (2001)

5. Haase, P., Stojanovic, L.: Consistent Evolution of OWL ontologies. In: Gómez-Pérez, A., Euzenat, J. (eds.) ESWC 2005. LNCS, vol. 3532, pp. 182–197. Springer, Heidelberg (2005)
6. Kotis, K., Vouros, G.A., Alonso, J.P.: HCOME: A Tool-Supported Methodology for Engineering Living Ontologies. In: Bussler, C.J., Tannen, V., Fundulaki, I. (eds.) SWDB 2004. LNCS, vol. 3372, pp. 155–166. Springer, Heidelberg (2005)
7. Stojanovic, L., Maedche, A., Motik, B., Stojanovic, N.: User-driven Ontology Evolution Management. In: Proceedings of the 13th International Conference on Knowledge Engineering and Knowledge Management (2002)
8. Sure, Y., Erdmann, M., Angele, J., Staab, S., Studer, R., Wenke, D.: Ontoedit: Collaborative ontology development for the semantic web. In: Proceedings of the 1st International Semantic Web Conference (2002)
9. Cimiano, P., Handschuh, S., Staab, S.: Towards the Self-Annotating Web. In: Proceedings of the 13th International World Wide Web Conference, WWW 2004, ACM Press, New York, USA (May 2004)
10. Cahier, J.-P., Zaher, L., -P., L.J., Pétard, X., Guittard, C.: Experimentation of a socially constructed "Topic Map" by the OSS community. In: Proceedings of the IJCAI 2005 workshop on Knowledge Management and Ontology Management, Edinburgh, August 1 (2005)
11. Cahier, J.-P., Zacklad, M.: Socio Semantic Web applications: towards a methodology based on the Theory of the Communities of Action. In: COOP 2004 Workshop on Knowledge Interaction and Knowledge Management (2004)
12. Vander Wal, T.: Folksonomy, Folksonomy Coinage and Definition. Vanderwal.net (February 2, 2007), http://vanderwal.net/folksonomy.html
13. Golder, S.A., Huberman, B.A.: The structure of collaborative tagging systems. Journal of Information Science 32(2), 198–208 (2006)
14. Kroski, E.: The Hive Mind: Folksonomies and User-Based Tagging. Infotangle (December 7, 2005)
15. Peterson, E.: Beneath the Metadata: Some Philosophical Problems with Folksonomy, D-Lib Magazine, 12(11) (November 2006),
    http://www.dlib.org/dlib/november06/peterson/11peterson.html
16. Gruber, T.: Ontology of Folksonomy: A Mash-up of Apples and Oranges (2005)
17. Lachica, R.: Qualitative semantic distance measurement test of tag relations (2007),
    http://www.hyposoft.no/papers/sem-dist-measure-test-social-bookm-tag-rel.pdf
18. Miller, G., Charles, W.G.: Contextual Correlates of Semantic Similarity. Language and Cognitive Processes 6(1), 1–28 (1991)
19. Guy, M., Tonkin, E.: Folksonomies: Tidying up tags? D-Lib Magazine, 12 (2006),
    http://www.dlib.org/dlib/january06/guy/01guy.html
20. Merholz, P.: Metadata for the masses. Adaptive path (October 19, 2004),
    http://www.adaptivepath.com/publications/essays/archives/000361.php
21. ISO/IEC stage 13250-2: Topic Maps — Data Model. In: International Organization for Standardization, Geneva, Switzerland, June 18, 2006 (2006), Available at
    http://www.isotopicmaps.org/sam/sam-model/2006-06-18
22. Fodor, J., Lepore, E.: Holism, A Shopper's Guide. Blackwell, Oxford (1992)
23. Preece, J.: Online Communities: Supporting Sociability, Designing Usability. John Wiley, Chichester, England (2000)
24. Hirsh, S.G., Dinkelacker, J.: Seeking Information in Order to Produce Information. Journal of the American Society for Information Science and Technology 55(9), 807–817 (2004)

25. Karabeg, D.: Polyscopic Modeling Definition. In: Griffin, R., et al. (eds.) Changing Tides. Selected Readings of IVLA (2004)
26. Barabási, A.L.: Linked: The new science of networks. Mass. Perseus Publ., Cambridge (2002)
27. Guescini, R.B., Karabeg, D., Nordeng, T.: A Case for Polyscopic Structuring of Information. In: Charting the Topic Maps Research and Applications Landscape, pp. 125–138. Springer, Heidelberg (2006)
28. Bloom, B.: A Taxonomy of Educational Objectives. In: Cognitive Domain. McKay (1965)
29. Beenen, G., Ling, K., Wang, X., Chang, K., Frankowski, D., Resnick, P.: Using social psychology to motivate contributions to online communities. In: CSCW 2004: Proceedings of the ACM Conference On Computer Supported Cooperative Work. ACM Press, New York (2004)
30. Shadbolt, N., Hall, W., Berners-Lee, T.: The Semantic Web Revisited. IEEE Intelligent Systems Journal, 96–101 (May/June, 2006)
31. Bourguin, G., Derycke, A., Tarby, J.C.: Beyond the Interface: Co-evolution inside Interactive Systems - A Proposal Founded on Activity Theory. In: Proc. of IHM-HCI (2001)
32. Morville, P.: Ambient Findability. O'Reilly Media. Sebastopol, CA (2005)
33. Kitcher, P.: Science, Truth, and Democracy. Oxford University Press, New York (2001)
34. Haase, P., Sure, Y.: State-of-the-art on ontology evolution, Institute AIFB, University of Karlsruhe, SEKT informaldeliverable 3.1.1.b (2004)

Ruby Topic Maps

Benjamin Bock

University of Leipzig, Johannisgasse 26, 04103 Leipzig, Germany
bb-lncs-tmra2007-rtm@bockb.de

Abstract. Ruby Topic Maps (RTM) is a Topic Maps engine created in
and for the Ruby programming language. Its focus is an intuitive and
easy to use interface, or (as the creators of Ruby would express it): RTM
aims to be the Topic Maps programmer's best friend. The development
of this library is a first step to enable Ruby programmers to work with
the Topic Maps technology. The library is considered to be useful on its
own but also as the basis for higher level tools, e.g. a Ruby on Rails
plug-in with an interface like ActiveRecord's.

1   Introduction

Ruby[1] and Ruby on Rails[2] in particular are resounded throughout the land
of Web 2.0 while Topic Maps is a standard for representation and interchange
of knowledge. This is a perfect match, but, until now, library support for Topic
Maps in Ruby was practically non-existent. RTM is a first step to fill this gap. It
aims to be a full-featured library for Topic Maps with an unobtrusive programming interface, extensive facilities to import from and export to common Topic
Maps formats and seamless integration with other Ruby libraries. Still, RTM
itself does not strive to provide higher level features, like topic-object mapping
or visualization; these features are not generic enough to belong to the engine
but will eventually be built on top.

There are design decisions to make to satisfy the dynamics and agility of
Ruby while respecting the strictness of the Topic Maps standards. "Convention
over Configuration" is a buzzword that describes this situation and this paper
proposes and explains the approach of Ruby Topic Maps.

2   Ruby, Its Ideals and Expectations to a Ruby Library

Ruby is an interpreted, object-oriented programming language, but also features
procedural and functional paradigms. In Ruby, everything is an object, there are
no primitive types and every expression returns a value. Objects have a type but
variables do not have a type. Ruby is a duck typing[3] language. The emphasis
is on how an object acts, not what type it is: If it walks, swims and quacks like
a duck, I call it a duck (and treat it like one). Duck typing can be regarded

L. Maicher and L.M. Garshol (Eds.): TMRA 2007, LNAI 4999, pp. 172–185, 2008.

as implicit interfaces, as opposed to explicit interfaces in other programming languages.

Ruby features on purpose single inheritance only. Instead of multiple inheritance, it has modules, similar to Categories in Objective-C, which are collections of methods. A module can be mixed into a class. In this case, the class receives all the methods of the mixed-in module, as if these methods were defined explicitly in the class' source code. Classes are generally open, so methods can be added, removed or modified at any time. A here-document syntax similar to Perl's or Bash's simplifies insertion of long strings, but for generation of HTML, embedded Ruby, which works like JSP or PHP, is more commonly used. Also, there are iterators, closures (based on passing blocks of code), continuations, generators and dependency injection.

There is a strong philosophy around the language and its libraries: the emphasis is on "human needs rather than computer needs", "the principle of least surprise" (POLS)[4], "don't repeat yourself" (DRY)[5] and others. Still, Ruby is not chaotic but very well structured: there are four levels of function scope (global, class, instance, and local), there is exception handling and automatic garbage collection.

The language has syntax support for Perl-like regular expressions (`/expr/`), arrays (`a = [1,2]`) and arbitrary finite mappings, called hashes (`{ "key" => "value" }`), build-in support for reflection, metaprogramming, and operator overloading. A method called `method_missing` receives invocations of methods which do not exist. An implementation of `method_missing` may also dynamically generate methods such that formerly missing methods are available for subsequent invocations.

The method names should be self-explanatory, which is supported by punctuation marks: Many invocations of methods which return a boolean value resemble asking a question. The names of such methods should end with a quotation mark. An example invocation is `object.valid?` in contrast to Java style `object.isValid()`. The names of destructive methods[1] should end with an exclamation mark, the names of setters should end with an equals sign, however getters should not end with any punctuation mark. Whitespace is allowed between the core method name and the equals sign, which[2] allows invocations of setters to look like normal variable assignment statements, for example `powerplant.boss = burns` instead of `powerplant.boss=(burns)`.

Symbols are basically words which start with a colon. Unlike strings, they are unique in memory and equal symbols have the same object id. They are often used as constants or identifiers, for example as keys in hashes. In the current version of Ruby, strings and symbols are not equal (thus `"example" != :example`), but they may be converted into each other.

---

[1] Destructive methods change the internal state of an object instead of returning a new altered object. E.g. `a=" text "; a.strip!` removes the whitespace and changes `a` while `a.strip` does not change `a`.

[2] Together with the possibility to omit the parenthesis for method invocation with one parameter.

3   Status Quo of Ruby Topic Maps

The idea for RTM was born in late 2006 when the author vainly searched for a Topic Maps engine in Ruby. He was inspired by the ease of use of many Ruby libraries and took up the work to implement a Topic Maps engine from scratch. Most of the development work was done in summer 2007 and a first preview version was released in October 2007 with a tutorial at TMRA. The work continues slowly but steadily.

RTM is an implementation of the Topic Maps - Data Model (TMDM)[6] and features an in-memory as well as a persistent database back-end. Both back-ends rely on the object-relational mapper ActiveRecord which is part of Ruby on Rails[2]. The RTM memory back-end makes use of the SQLite `:memory:` back-end while the persistent back-end supports all databases supported by ActiveRecord[3].

RTM is developed using Model-Driven Development (MDD) techniques. The TMDM is implemented in a custom Domain-Specific Language (DSL) which describes all TMDM items including their properties, equality rules, constraints, and the rules for merging. It also includes the core relationships defined by the TMDM, namely type-instance and supertype-subtype. These core relationships are reflected by the methods `types`, `instances`, `supertypes`, and `subtypes`. Each method returns a set of topics matching the corresponding relationship.

The back-end is an interpreter of this DSL which generates the actual source code. The TMDM standard suggests implementing TMDM items as classes. Their names are obvious: class `Topic` for topic items, class `TopicName` for topic name items and so on. Accordingly, the properties of the items are instance methods of the classes. The API sticks strictly with the terms used in the TMDM. Each property becomes a getter with exactly the name from TMDM, e.g. `TopicMap` has a getter `topics`. This way, users who are new to Topic Maps can start coding using only the TMDM specification without having to look at the documentation all the time.

Additionally to the TMDM naming, also Java TMAPI[7] naming is supported. E.g. the property "parent" of an association role can be accessed by both TMDM-style `parent` and TMAPI-style `association`[4].

If applicable, there is also a setter for the property. The name of a setter is the name of the property with an equal sign appended. This means, the Java TMAPI code `topicname.setValue("foo")` becomes `topicname.value = "foo"` and, similarily, `role.getAssociation()` becomes `role.association`. The details of the API will be discussed below.

The two back-ends available right now share the bigger part of their code, because the memory back-end also uses the object-relational mapper ActiveRecord.

---

[3] Databases currently fully supported: MySQL, PostgreSQL, SQLite, SQL Server, Sybase, and Oracle. For DB2, the schema is not created automatically by RTM but must be inserted by the DBA because ActiveRecord::Migration does not support DB2 yet.

[4] In Java it is actually called `getAssociation()` which translates to `association` in Ruby-style.

From the development perspective, differences are mostly in the connection code and schema generation. From the user's perspective, of course, the performance is different and all data is discarded when the application terminates.

ActiveRecord is the model part in the model-view-controller architecture of Ruby on Rails, but it can also be used without Ruby on Rails. ActiveRecord is the most widespread Ruby implementation of the Active Record design pattern[8].

In RTM, the database schema is defined using an ActiveRecord Migration, a module which produces database-specific SQL statements from a database-agnostic schema definition in Ruby code. The database is not accessed directly by the RTM code but through the ActiveRecord model classes. The benefits are an enormous ease of use and a vast amount of functionality just by the definition of the basic relations between the classes. These benefits come at the cost of restricted flexibility. Coupling RTM too tightly to ActiveRecord would make RTM dependent on the underlying database type and database schema, something which should be avoided for a stable API.

RTM solves this by adding a thin wrapper layer on top of the ActiveRecord classes. As mentioned above, this layer is based on the MDD definition. The architecture allows it to develop other back-ends on top of existing persistence frameworks, e.g. a REST[9] web service back-end on top of Active Resource, which is the REST back-end of Ruby on Rails.

The natively supported serialization format is XTM 2.0[10]. Compact Topic Maps (CTM)[11], Linear Topic Maps (LTM)[12], JSON Topic Maps (JTM)[13], and YAML Topic Maps (YTM) are still in the works and only partially supported yet. Additional formats can be added and will seamlessly integrate into the API.

4   Usage of Ruby Topic Maps

4.1   Installation

Provided a recent Ruby default setup is already installed, the installation of Ruby Topic Maps is just one step, thanks to Ruby's package management system RubyGems. As a system administrator type

```
gem install rubytopicmaps
```

For special configurations like single user installation refer to the RubyGems manual.

In the special case of RubyGems not being available, the release can be downloaded at the RTM project homepage[5].

Instead of installing RTM systemwide, RTM can also be made available directly to a particular Ruby project. This can be accomplished by copying the RTM source directly into the lib folder of that project or by adding the path to the RTM source to the internal library path:

```
>> $: << "/path/to/custom/sources/"
>> require 'rtm'
```

---

[5] http://rtm.rubyforge.org/

4.2   Hello Topic Maps World

For the first steps, the command line tool `irb` is a good help. Everything entered there can also be written in a real Ruby program and vice versa. It is started using `irb` in the command prompt. The optional parameter `--simple-prompt` gives a shorter prompt like the one used in the examples here. The input is preceded by `>>` and the debug output is preceded by `=>`. To use RTM, it is necessary to load the library. The package is internally split into several modules which can be loaded individually. For simplicity there is a module which loads all common modules in one step.

```
>> require 'rtm'
=> true
```

Because `irb` is interactive, it prints the result of every single command. In the following examples, the output is often shortened to keep the text readable.

The next step is to connect to a back-end. The easiest way is to use the memory back-end, as it works without any additional configuration:

```
>> RTM.connect
=> true
```

To connect to other back-ends, the corresponding parameters must be passed to the `connect` method. For the ActiveRecord back-end, the parameters to the `connect` method are equivalent to the parameters of the `establish_connection` method in ActiveRecord and documented there. Common uses are covered by some wrapper methods. The following example illustrates the connection to a MySQL database using the ActiveRecord back-end. Of course, the database schema needs to be applied to the database before the first use. This can be done directly in the code, for example in a setup script of an application. For the memory back-end, this is not needed. Finally, the first topic map can be created.

```
>> RTM.connect_mysql("database", "username", "password", "host")
=> true
>> RTM.generate_database
=> true
>> tm = RTM.create "http://rtm.rubyforge.org/examples/1"
=> #<RTM::TopicMap...>
```

The String parameter of the `create` method is the locator for the topic map. The topic map created above can now be accessed using the local variable `tm` or any other variable it was assigned to. The topic map could also be fetched again using the internal numerical id

```
>> tm = RTM[0]
=> #<RTM::TopicMap...>
```

or using its locator

```
>> tm = RTM["http://rtm.rubyforge.org/examples/1"]
=> #<RTM::TopicMap...>
```

4.3   Creation of Topic Maps Constructs

New topics can be created with the command `tm.create_topic`, new associations with `tm.create_association`. All Topic Maps constructs can be created by the corresponding `create_thing` method in their parent, the same way it is done in TMAPI. The arguments to the method are the properties which are necessary to create the Topic Maps construct according to the restrictions of the TMDM. In TMAPI, the positional arguments are mandatory. Not so in RTM: The necessary properties may also be set using keyword parameters or in the block.

To clarify that, here is a short excursion to method invocation in Ruby. The formal definition of the parameter list for a method is the following:

```
parameters := ( [param]* [, hashlist] [*array] [&aProc] )
```

A list of arbitrary length of positional parameters may be followed by a `Hash`. Ruby does not directly support keyword parameters but they can be emulated using this `Hash`. A `Hash` expression is normally surrounded by braces, but if it is unambiguous, the braces may be omitted. A method invocation with one (emulated) keyword parameter is given in the next example:

```
o.my_method :keyword => "parameter"
```

The third type of parameter is the so-called splat-array. Instead of multiple method definitions, optional parameters are passed to the single implementation of the method in form of an array. The last type of parameter is a `Proc` object. `Procs` are blocks of code, also known as closures, which have been bound to a set of local variables. In RTM, they are mainly used to work with newly created objects and for querying and filtering. Examples of usage are given below.

When parsing input, e.g. from a HTTP request, the result is usually processed to a `Hash`. Instead of reading the objects from the `Hash` and passing them as positional parameters to a `create_thing` method, the `Hash` can be directly passed.

```
>> at2, p2, rt2 = tm.create_topic, tm.create_topic, tm.ct
=> [#<RTM::Topic...>, #<RTM::Topic...>, #<RTM::Topic...>]
>> a2 = tm.create_association at2
=> #<RTM::Association...>
>> role3 = a2.create_role :type => t2, :player => p2
=> #<RTM::AssociationRole...>
```

If a `Hash` given at a method invocation assigns a value to a symbol, while the symbol matches the name of a positional parameter variable in the declaration of the method, the value from the Hash will be used instead of the value given as positional parameter. Thus, in the following example, "name1" will be discarded and "name2" will be used.

```
>> topic.create_name "name1", :value => "name2"
=> #<RTM::TopicName...>
```

Often, a newly created object will be used (and modified) in the source code below. To address this, a block may be passed to **create_*thing*** methods. The block receives one parameter, the newly created object, which may not be persisted yet. This technique may be used to optimize commits in the back-end. A specialized back-end could treat the block as transaction which is committed after executing the block. The example shows the creation of a topic within the `TopicMap` instance `tm`. The block is surrounded with **do** and **end** and the parameter is enclosed in vertical bar characters:

```
>> tm.create_topic do |topic|
>>    topic.create_name ...
>> end
=> #<RTM::Topic...>
```

The following example creates two new topics, an association and a role for this association with the topics assigned as player and type for the role. The association is not yet valid at the beginning of the block as it has no roles. At the end of the block, marked by **end**, the association is valid and gets committed.

```
>> player1 = tm.create_topic "player1"
=> #<RTM::Topic...>
>> roletype1 = tm.create_topic "roletype1"
=> #<RTM::Topic...>
>> assoc1 = tm.create_association "assoctype1" do |a|
>>    role1 = a.create_role player1, roletype1
>> end
=> #<RTM::Association...>
```

4.4 Referencing Topics

In the majority of cases, the arguments passed to methods of the library are Topic Maps objects, or, technically, references to objects. This makes it necessary to look up objects by some identifying property before using them. For topic object references, RTM provides a shortcut to alleviate the need to write a separate lookup statement: Wherever a topic object reference is required, supplying a string which identifies the mentioned topic is sufficient. There are three ways of addressing a topic and consequently three additional ways to reference a topic in RTM:

1. reference by item identifier – provide a relative IRI[6] which will be resolved against the locator of the topic map.
2. reference by subject identifier – provide an absolute IRI or a QName
3. reference by subject locator – provide an absolute IRI or a QName prepended by "="

---

[6] Item identifiers are always absolute IRIs in TMDM. To look up a topic by an absolute item identifier use the method `tm.topic_by_item_identifier`. This method works with both, relative and absolute IRIs.

Before a QName is used, its prefix must be registered. The example shows the (re-)definition of the predefined prefix for XML Schema and the definition of a custom prefix:

```
RTM.prefix "xs", "http://www.w3.org/2001/XMLSchema#"
RTM.prefix "custom", "http://rtm.rubyforge.org/exampleprefix"
```

To create a new topic with a subject identifier representing the XML Schema type "integer", the following code may be used:

```
>> tm.create_topic "xs:integer"
=> #<RTM::Topic...>
```

To get a specific topic object from a reference, the method `tm.get` is used. In the following example, `"topic_reference"` may be a string in one of the three ways described above.

```
t = tm.get("topic_reference")
```

If the topic does not exist, TODO `nil` is returned. The method `tm.get!` works in a similar way, but it creates the topic if it does not exist.

```
t = tm.get!("topic_reference")
```

The next chapter additionally shows how to use topic references instead of topic objects when creating associations and association roles.

4.5 Shortcuts

While developing an application it is useful from time to time to use the `irb` command line to just try something quickly. Typing the long method names again and again can cost a lot of time, even though `irb` has a completion feature. That is why RTM provides shortcuts for the most commonly used methods. As a rule of thumb, all sets of objects can be abbreviated using just their first letter and all create_*thing* methods can be abbreviated as c*t*, the first letter of create and the first letter of the object to create.

```
>> tm.topics[0].names[0].value
```

becomes

```
>> tm.t[0].n[0].value
```

Taking shortcuts and automatically generated, string referenced topics together, a topic map with six topics, three roles, and one association is created in a few seconds of typing using just Ruby code.

```
>> tm.ca "parents-child" do |a|
>>    a.cr "Homer", "parent"
>>    a.cr "Marge", "parent"
>>    a.cr "Lisa", "child"
>> end
```

The complete list of shortcuts can be found in the documentation and will not be duplicated here.

4.6 Import

The most general way to import a topic map is using the `RTM.from` method. This method takes the source in various forms and tries to detect what type of data format the source provides. `RTM.from` accepts `File`, `URI`, and `IO` objects. It downloads and opens them if necessary and delegates the work to one of the specialized functions for the source format. It is also possible to pass a `String` which will be examined if it is a path to an existing file or a valid URI and treated as described above or attempted to be parsed directly. Some examples of usage,

```
tm2 = RTM.from "/home/bb/topicmaps/example2.xtm2"
tm3 = RTM.from "http://rtm.rubyforge.org/example3.ltm"
```

If not specified separately, the document locator is the path given. This behaviour may be overridden using the keyword parameter `:document_locator`

```
tm4 = RTM.from "example4.ctm",
               :document_locator => "http://example.com/tm4"
```

The destination topic map is created using the document locator unless specified as the `:topic_map` parameter.

```
    RTM.from "example5.ctm",
             :document_locator => "http://example.com/tm5",
             :topic_map => tm4
```

The other parameters can be specified in this way too: `:source` specifies the source which was given as the first parameter in the previous examples. There is also `:format` which specifies the content type (and disables content type detection) as well as `:type` which sets the type of source, for example `:string`, to tell the `from` method not to attempt to open it as a file.

With regard to opening web resources, HTTPS is supported and user credentials for HTTP Basic Authentication[14] can be supplied by the parameters hash as described in the documentation for Ruby standard library, package `OpenURI`[7]. The data gets directly streamed to the import function, which may update the destination topic map even before finishing the download. Other options may vary between the different import plug-ins.

```
tm5 = RTM.from :type => :string, :format => :ltm, :source => <<EOF
  parents-child(Homer : parent, Marge : parent, Lisa : child)
EOF
```

The result is a topic map which is equivalent to the hard-coded example of the chapter Shortcuts above.

To implement a new import format, only the method `from_formatname` must be implemented and must accept an `IO` object. The functionality for opening files and URIs is added by RTM.

---

[7] http://www.ruby-doc.org/core/classes/OpenURI/OpenRead.html

The most significant import format is probably XTM 2.0. The current implementation uses the library REXML, which is written in plain Ruby and part of the Ruby standard library. REXML provides a DOM tree interface and a SAX-style parsing interface. While the former regards the XML document as a random-accessible object structure, the latter regards the XML document as a series of events. The DOM tree interface is more convenient to work with, but it needs main memory roughly proportional to the size of the XML document. The SAX-style parsing interface only needs main memory roughly proportional to the nesting depth of the XML document, and thus it scales considerably better. The XTM 2.0 importer of RTM uses the SAX-style interface of REXML.

4.7 Export

Export works analogously to import, aligned with the Ruby naming convention the method name prefix is called `to_`. This means the method names are constructed by concatenating `to_` and the short name of the destination format together. Thus, an export to a string in the format JTM can be accomplished by `tm.to_jtm`. If one parameter is supplied, the export methods accept a string (indicating the name of the destination file) or an IO object (indicating the destination stream). If no parameter is supplied, the export methods return the serialized representation as a string.

For the RTM objects from TMDM, the standard Ruby method `to_s` (to string) takes an optional format symbol, for example `:short` and `:long` to control the formatting style of the output. The `:long` output can be really handy while debugging.

4.8 Querying and Filtering

Thanks to Ruby's `Enumerable` module, which is mixed into all sets of constructs from TMDM, a simple way querying is built in the language itself. The concept of `Enumerable` is to provide several traversal and searching methods to collection classes. The programmer of the collection class implements the iterator function `each` which successively yields the members of the collection. A code block can be passed to most of the methods from `Enumerable`. This block is evaluated for each object in the collection. The code in the block may for example transform a single object or express the condition to be checked for a single object. The most important methods of `Enumerable` will be presented here, the complete list is part of the Ruby Documentation[8].

The query for a specific object starts at a set which holds this kind of objects. Most times, this will be the set `tm.topics` within a topic map `tm`. The current implementation evaluates the condition given in the block in memory only. This means: every Topic Maps object must be fetched from the database (or another back-end) in order to evaluate its properties. For small examples, this works fine. But for big deployments, this may be a bottleneck.

---

[8] http://ruby-doc.org/core/classes/Enumerable.html

To address this problem, which also exists for other libraries, the Ruby library ParseTree[9] was created. This library helps to parse the Ruby statements (i.e. conditions) given in the block. It is the basis for the Ambition Project[10] by Chris Wanstrath. Ambition is a framework which allows to write adapters which transform simple Ruby code into any other API. This will be further evaluated to optimize querying in a later version of RTM.

Until that, there are two options: Accepting the overhead and using the in-memory queries or using back-end-specific queries, i.e. ActiveRecord for the back-end described here. Both will be outlined below.

The following example returns a list of all topics which have no name:

```
>> no_names = tm.topics.find_all { |t| t.names.size == 0 }
=> [#<RTM::Topic...>, #<RTM::Topic...>, ...]
```

The method find_all works on the set topics and checks for each topic t in the set whether the condition in the block evaluates to true. The first part of the block is the argument list, surrounded by vertical bars. The second is a list of statements. In the given example, |t| is the argument list and the statement refers to the topic t. The statement counts the names of t and compares the result to zero. The method find_all returns all topic objects t for which the statement in the block evaluates to true.

To find a topic[11] which has an occurrence with the value "10", the next snippet can be used. For brevity, the shortcuts described above are used.

```
>> to10 = tm.t.find{|t| t.o.select{|o| o.value == "10"}.size > 0}
=> #<RTM::Topic...>
```

This is how it works: all topics in the topic map are iterated. For each topic t, all the occurrences which have the value "10" are selected and counted. When the first topic which has at least one occurrence with the value "10" is found, this topic is returned. The above example is used to illustrate the possibilities. A much more efficient way to query this would be:

```
>> to10 = tm.occurrences_by_value("10").map { |o| o.parent}
=> [#<RTM::Topic...>, #<RTM::Topic...>, ...]
```

All occurrences (within all topics) in the topic map tm will be searched for the value "10" and then, for each matching occurrence, the parent object, which is a topic, will be returned.

To clean up a topic map, all topics which do not play a single role shall be deleted. But first, we want to know if there are any.

```
>> tm.topics.any? { |t| t.roles.size == 0 } #returns true or false
=> true
>> tm.topics.select { |t| t.roles.size == 0}.each { |t| t.remove }
```

---

[9] http://parsetree.rubyforge.org/
[10] http://ambition.rubyforge.org/
[11] In case there are more than one matching topics, an arbitrarily chosen topic of the set of matching topics will be returned.

To output a comma-separated list of all names of a topic, line-by-line for each topic, the following may be helpful:

```
>> tm.t.each { |t| puts t.names.map { |n| n.value }.join(", ") }
```

Please note that, when printing a name directly, it is not necessary to append the .value to the name object, because the name's value is the default return value of the to_s method of each name object. However, in the example above, the .value or an explicit .to_s is needed, as otherwise join would not be invoked on a list of string objects, but on a list of name objects, which do not support this type of joining.

The ActiveRecord queries are uniformly accessed via the find methods of the sets to query and provide only the raw ActiveRecord functionality without additional Topic Maps knowledge. The following example shows querying all associations of type type-instance[12]. For frequently used Published Subject Identifiers, RTM provides shortcuts, e.g. RTM::PSI[:type_instance]. The result set is a list of wrapped association objects which can be used as usual.

```
>> tid = tm.get(RTM::PSI[:type_instance]).id
=> 10
>> tm.associations.find(:all, :conditions => {:ttype_id => tid})
=> [#<RTM::Association...>, #<RTM::Association...>, ...]
```

This example shows two drawbacks of this approach: The internal ids have to be fetched and compared to the corresponding database columns. This requires in-depth knowledge of the underlying database schema, but currently, this way of querying offers by far the best available query speed in RTM.

5   Performance and Scalability

The initial focus of the development was to create an easy to use programming interface. For this first development milestone, runtime performance was not a requirement. Due to Ruby's interpreted and dynamic nature, a Topic Map engine written in Ruby is presumably slower than a Topic Map engine written in compiled languages. However, there are several starting-points to widen bottlenecks.

The approach of query optimization was already mentioned above. Topic objects are queried often, this also provides much potential for caching. The get method of a topic map is a central point of access where caching will be implemented.

Database views which each represent associations of common type, such as type-instance or supertype-subtype, can be created or even be automatically generated from runtime information. Another approach is joining associations with tuples of their association roles into one (cached) view table with the column schema:

```
player0,role0,association_type,role1,player1
```

---

[12] The complete IRI for this PSI is http://psi.topicmaps.org/iso13250/model/type-instance

Such a representation alleviates the need of changes to the database (i.e. creating views) for each new cached association type or role type and covers all unary and binary associations completely.[13]

All of the above can happen transparently for the programmer who uses RTM. An application can directly benefit from optimizations in the RTM core library.

The speed of RTM depends heavily on the speed of the underlying database system. Database comparison tables and charts with current results can be found at the RTM Homepage[14].

6   Further Development

This new engine has yet to prove itself in everyday programming life. The unit tests must become more comprehensive, and, probably, there will be bugs to be fixed and issues to be resolved. New import and export formats will be added and new back-ends could be developed. An interesting option is to integrate RTM with any Java TMAPI implementation using JRuby.

Along this way of stabilization, the work continues at a higher level. Active-TopicMaps, a Ruby on Rails plug-in, will provide an alternative model to, and should become the fellow of, ActiveRecord. It strives to be as easy as using ActiveRecord Migrations and ActiveRecord Associations to describe an application's data model. On the other hand, it will integrate with ActionController, to allow Ruby on Rails Scaffolding, which is the automatic generation of HTML forms with view and editing capabilities.

References

1. Ruby Community: Ruby, http://www.ruby-lang.org/
2. Hansson, D.H.: Ruby On Rails, 37 Signals, http://rubyonrails.com/
3. Alex Martelli in a message to the comp.lang.python newsgroup (July 26, 2000), http://groups.google.com/group/comp.lang.python/msg/e230ca916be58835
4. Raymond, E.S.: Applying the principle of least surprise. The Art of Unix Programming, http://www.faqs.org/docs/artu/ch11s01.html
5. Hunt, A., Thomas, D.: The Pragmatic Programmer. In: The Evils of Duplication, Ch. 7. Addison-Wesley (1999), http://www.pragmaticprogrammer.com/ppbook/extracts/rule_list.html
6. ISO/IEC IS 01325-2:2006: Information Technology – Document Description and Processing Languages – Topic Maps – Data Model. International Organization for Standardization, Geneva, Switzerland (2006), http://www.isotopicmaps.org/sam/sam-model/
7. Ahmed, K., Garshol, L.M., Grønmo, G.O., Heuer, L., Lischke, S., Moore, G.: Common Topic Map Application Programming Interface (2004), http://www.tmapi.org/

---

[13] Acknowledgement: The optimization considerations in this paragraph have been developed jointly by the author and Xuân Baldauf.

[14] http://rtm.rubyforge.org/docs/tuning/

8. Fowler, M.: Patterns of Enterprise Application Architecture. Addison-Wesley, Reading (2002)
9. Fielding, R.T.: Architectural Styles and the Design of Network-based Software Architectures. PhD Dissertation, University of California, Irvine (2000), `http://www.ics.uci.edu/~fielding/pubs/dissertation/top.htm`
10. ISO/IEC IS 13250-3:2007:: Information Technology – Document Description and Processing Languages – Topic Maps – XML Syntax. International Organization for Standardization, Geneva, Switzerland (2007), `http://www.isotopicmaps.org/sam/sam-xtm/`
11. ISO/IEC Draft 13250-6:2007: Information Technology – Document Description and Processing Languages – Topic Maps – Compact Syntax, 2007-11-16. International Organization for Standardization, Geneva, Switzerland (2007), `http://www.isotopicmaps.org/ctm/`
12. Garshol, L.M.: The Linear Topic Map Notation, Version 1.3 (June 17, 2006), `http://www.ontopia.net/download/ltm.html`
13. Cerny, R.: Topincs - A RESTful Web Service Interface for Topic Maps. In: Maicher, L., Sigel, A., Garshol, L.M. (eds.) TMRA 2006. LNCS (LNAI), vol. 4438, pp. 175–183. Springer, Heidelberg (2007)
14. Franks, et al.: HTTP Authentication: Basic and Digest Access Authentication, RFC, IETF 2617 (June 1999), `http://www.ietf.org/rfc/rfc2617.txt`

Expressing Dublin Core in Topic Maps

Steve Pepper

Ontopedia
Oslo, Norway
pepper.steve@gmail.com

Abstract. The Dublin Core Metadata Initiative is an open organization engaged in the development of interoperable online metadata standards that support a broad range of purposes and business models. Its most important standard is the Dublin Core Metadata Element Set [1], a vocabulary of fifteen properties for use in resource description, which was approved as ISO 15836:2003 [2]. This and other vocabularies developed by Dublin Core are defined as abstract models which may be expressed in any number of different syntaxes. This paper presents a proposal for expressing such metadata using Topic Maps.

Keywords: metadata, Dublin Core.

1 Introduction

The Dublin Core Metadata Initiative (DCMI) is an "open organization engaged in the development of interoperable online metadata standards that support a broad range of purposes and business models." It has developed a number of metadata specifications, the most important and widely used of which is the *Dublin Core Metadata Element Set* [1], which was approved as an ISO standard in 2003 [2].

DCMES and three other vocabularies are documented in [3]. DCMES itself consists of 15 elements: contributor, coverage, creator, date, description, format, identifier, language, publisher, relation, rights, source, subject, title, and type. These are supplemented by 40 additional *Elements and element refinements*. There is also a vocabulary of 18 *Encoding schemes*, one of which (the *DCMI Type Vocabulary*) is also a Dublin Core (DC) vocabulary consisting of 12 terms.

In the DCMI abstract model [4], "resources" are "described" by metadata using "property-value" pairs. 'Resource' is defined as "anything that might be identified" and, although the concept of "being identified" is not defined, it is clear from the examples that 'resource' is essentially the same as 'subject' in Topic Maps. The resources to which Dublin Core metadata is assigned are almost always *information resources*, but this need not be the case; they might, for example, be "human beings, corporations, concepts [or] bound books in a library."

In terms of the abstract model, the first two vocabularies mentioned above define *properties*; the encoding schemes identify controlled vocabularies such as ISO 3166, TGN and DCMI Box, that provide *values* for properties such as **language**, **coverage**, etc.; and the DCMI Type Vocabulary defines a set of possible *values* (such as Text, Service, Image, etc.) for the property **type**. Thus, two vocabularies define properties, one defines schemes that are the source of values, and the fourth consists of actual values.

L. Maicher and L.M. Garshol (Eds.): TMRA 2007, LNAI 4999, pp. 186–197, 2008.

DC vocabularies are defined as abstract structures; no particular syntactic representation is mandated. Instead, metadata can be expressed in any syntax that is convenient to the user. Recommendations already exist for expressing Dublin Core in HTML [5], XML [6], and HTML [7], and a new work item was recently approved by ISO for a Type 3 Technical Report on *Expressing Dublin Core Metadata Using Topic Maps* [8].

There are a number of reasons why it would be useful to be able to express Dublin Core in Topic Maps: (1) in order to add metadata to individual topic maps; (2) in order to use Topic Maps to collate DC metadata; and (3) in order to exploit the merging capabilities of Topic Maps to combine DC metadata with other non-DC data. The advantages derived from each of these applications increases greatly if the same conventions are used in every case.

This paper presents an overview of previous work in this area, discusses the issues raised, and proposes a unified approach suitable for standardization. Feedback from the community is welcomed.[1]

2 Review of Previous Work

A number of proposals have been put forward over the years regarding how to use DC with Topic Maps. This section summarizes and compares them.

Algermissen 2001 [9] was the first published (but unfinished) attempt to use Dublin Core in a Topic Maps context. It is described as "a processing model for HTML <meta> tags that make use of the Dublin Core element set" and consists of a set of PSIs in the form of a topic map modeled on psi1.xtm[2] within the framework of the Topic Maps Processing Model (TMPM4).

The topic map defines TMPM4 association templates that model 11 of the 15 core DC metadata elements. The element **creator** is typical: topics represent the resource and the value, and they are related via an association whose type corresponds to the metadata element ('creator'). Association types and role types are assigned PSIs in the *http://www.topicmapping.com/psi/dc/dc.html#* namespace; role types also have the DC identifiers as additional subject identifiers.

The following code shows how a **creator** property would be expressed:[3]

```
dcc:at-resource-creator(
dcc:role-resource : the-document,
dcc:role-creator : the-creator )
```

Nine other elements are treated in essentially the same way, the exception being **relation**, which is modeled as an "untyped relationship between resources". The remaining four elements are modeled as follows:

---

[1] Note: The goal of the current work is to standardize the usage of all four vocabularies documented in [3].

[2] psi1.xtm was the set of XTM 1.0 Published Subject Indicators (PSIs) published in December 2000 by TopicMaps.Org but subsequently withdrawn. It can be retrieved via the WayBack Machine at web.archive.org as http://www.topicmaps.org/xtm/1.0/psi1.xtm.

[3] The notation used for examples is the Compact Topic Maps syntax, as defined in the draft of 2007-09-09 [10]. The syntax is not yet stable and changes are likely.

- **title:** an untyped base name is assigned to the topic that represents the resource
- **description:** an occurrence of type 'description' is assigned to a topic that represents the resource
- **subject:** an untyped occurrence whose value is the resource is assigned to a topic that represents the subject
- **identifier:** [this element is marked as "to be done" with the comment "not sure if and how to handle this"]

The model requires 23 published subject identifiers in order to cater for 12 of the 15 core elements (**title** does not require additional PSIs since it is modeled as a base name, and the treatment of **subject** and **identifier** is incomplete).

To summarize: the Algermissen proposal uses names, associations and occurrences to handle 14 of the 15 core elements; associations are used in two different ways (typed, as in **creator**, and untyped, as in **relation**); occurrences are also used in two different ways (the resource "owns" the occurrence in the case of **description**, whereas the value "owns" it in the case of **subject**).

Pepper 2003 [11], reproduced in Appendix A, is an example of an approach pioneered by Ontopia and distributed as the file dc.ltmm with the *Omnigator* browser. It can be summarized as follows:

- **title** is represented as an untyped topic name
- **description**, **date** and **rights** are represented as occurrences
- all other core elements (except **identifier**) are represented as associations
- there is no example for **identifier** so its intended use is unclear
- the DC identifiers (e.g. *http://purl.org/dc/elements/1.1/creator*) are used as subject identifiers for occurrence types and association types
- PSIs are defined for the concepts 'resource' and 'value', respectively (in the namespace *http://psi.ontopia.net/metadata/#*) for use as role types.

Maicher 2006 [12] represents a similar approach, also published in the form of a topic map. It has the advantage of covering all four sets of terms documented in [3], but since it lacks documentation and complete examples, there is no way of knowing exactly how it is intended to work in practice. The two short examples in the topic map (showing metadata assigned to the topic map itself, and to the topic representing the subject "Der Schrei") include the following mappings:

- associations: **language, type, source**
- occurrences: **description, publisher, references, title**

The examples given for **creator** and **created** (the latter an "element refinement" of **date**) indicate that they can be treated as *either* occurrences *or* associations. Based on this one can assume that the intention is for users to be able to choose whether to use an occurrence or an association in any given instance.

Maicher follows Pepper in using the DC identifiers as subject identifiers, but also defines a parallel set (in the *http://psi.semports.org/dc/* namespace). He also follows Pepper in defining two role types to be used across the board whenever a metadata element is represented as an association; these are called 'start' and 'end'. No indication is given of how to handle the **identifier** metadata element.

3 Discussion

Assigning metadata to resources is equivalent to making statements about topics. It is therefore natural to represent the assignment of property-value pairs as statements of various kinds. In the case of the two DC vocabularies that define metadata elements (and element refinements) the key issue is to decide whether to represent a given property as a name, association, or occurrence.[4] This section discusses the issues involved before outlining a proposal for standardization.

3.1 Simplicity vs. Naturalness

The simplest possible solution would be to represent all DC metadata property-value pairs as occurrences, the value of which would be either a string or a URI. While this is eminently feasible, it would not produce results in keeping with the Topic Maps way of thinking.

One of the causes of the findability problem (which Topic Maps, of course, exists to solve) is the fact that much existing metadata consists simply of string values, which are subject to the well-known problems of homonymy and synonymy. In order to facilitate the aggregation of information, the Topic Maps paradigm encourages explicit reference to subjects via topics, rather than implicit reference via the strings that are used to name them: in other words, the approach is *subject-centric*. Values of properties should therefore be represented as topics wherever possible, and this means using associations rather than occurrences to represent property-value pairs.

Certain properties, however, are such that it would be counter-intuitive to represent them as associations.[5] An obvious example is **description**, which in Dublin Core typically consists of a sentence or two of text. It would engender unnecessary overhead if this were to be represented as a topic; a more natural solution is to use an occurrence. Another example is **title**, which is clearly a naming property, and is therefore most naturally represented as a topic name.

The only advantage of a "simpler" scheme based solely on occurrences is that it is easier to explain; its disadvantage is that it is less natural and leads to less useful results, and it is therefore to be rejected. The current proposal adopts the policy of using names, associations and occurrences as appropriate.

3.2 A More Flexible Scheme

A second issue is whether to mandate the use of one (and only one) kind of statement for any given metadata element, or whether to allow some degree of flexibility, perhaps coupled with certain admonitions concerning best practice. Algermissen and Pepper adopt the former approach, and Maicher the latter.

The most obvious example of this question arising is with elements which should ideally be represented as associations, but where it might sometimes be inconvenient to represent the value of the property as a topic. Many elements carry comments

---

[4] Handling the two vocabularies that define values and schemes for encoding values is trivial since terms belonging to these must clearly be mapped to topics.

[5] At least using the Topic Maps Data Model. It would be less counter-intuitive at the level of the Topic Maps Reference Model, which lacks the explicit semantics of the TMDM.

stating that "recommended best practice is to use a controlled vocabulary" (in which case an association, with the property value represented as a topic, would clearly be most appropriate). However this advice may sometimes not be followed and the value of the property in question might simply be entered as a string. In such cases one might question the utility of creating a topic about which nothing is known except its name: the situation is scarcely better than using a metadata property whose value is a string.

In our opinion, the importance of encouraging best practice outweighs this. By requiring the use of associations in such cases, a point is made about the "right way" to do things, and this will hopefully lead to the creation of more maximally useful topic maps. The current proposal is therefore prescribes exactly which kind of statement to use for each element, and mandates the use of associations in all but exceptional cases.

3.3 Role Types

Given the fact that associations are recommended for most elements (9 out of 15 in the basic element set), it becomes necessary to define appropriate role types. While it would be possible to define a different pair of role types for each association type, there seems to be no good reason to do so.[6] Following the principle of Occam's razor,[7] the current proposal recommends using the same pair of role types for every association type. These obviously need to be very generic; following Pepper (2003), the subjects 'resource' and 'value' are proposed, since these correspond directly to Dublin Core terminology:

```
%prefix dctm http://psi.topicmaps.org/iso29111/
dctm:resource - "Resource" .
dctm:value    - "Value" .
```

3.4 Subject Identifiers

As we have seen, both Algermissen (2001) and Maicher (2006) create new PSIs for every DC element (in addition to reusing the DC identifier), whereas Pepper (2003) simply reuses the existing identifiers. In the absence of any stated reason for the duplication it seems sensible to again apply Occam's razor and avoid creating new subjects unnecessarily. The current proposal therefore reuses the DC identifiers without creating new ones (except for the two role types described above).

3.5 The *Identifier* Element

All previous proposals have been unclear on the handling of the **identifier** element, whose DC definition is "an unambiguous reference to the resource within a given context". The accompanying comment states that "recommended best practice is to

---

[6] There are some who regard the reuse of role types across association types as bad practice. They are invited to state their objections.

[7] *Entia non sunt multiplicanda praeter necessitatem* (entities should not be multiplied unnecessarily).

identify the resource by means of a string conforming to a formal identification system." Most identifiers tend to be URIs, but it is important to note that they need not be.

Now, since the value of an occurrence can be a string, the use of occurrences to represent **identifier** elements would provide a general solution. However, since Topic Maps has two additional constructs that carry identification semantics – subject identifiers and subject locators – it is necessary to investigate the possibility of using these instead.

Both of the latter are defined as 'locators' in the Topic Maps standard, and locators are simply strings that conform to some locator notation (note that they need not be URIs). This means that, from the point of view of their syntax and data type, either subject identifiers or subject locators could be used to represent **identifier** elements. The key question is whether this would convey the correct semantics. To answer this question, it is necessary to consider the following definitions:

- **subject identifier:** a locator that refers to [an information resource] in an attempt to unambiguously identify [a subject]
- **subject locator:** a locator that refers to the information resource that *is* the subject of a topic

A subject identifier involves *indirect* identification: it locates an information resource that is *different from* the subject that it identifies. A subject locator involves *direct* identification: it both locates and identifies the same information resource.

The question to be answered is whether one or other of these conditions can be assumed to exist *in every case* in which **identifier** is used. The answer is clearly no: The resource to which DC metadata is assigned is very often an information resource and its URL is often the value of its **identifier** property; in such cases it would be appropriate to use subject locators. However, a resource *need not* be network retrievable (it could, for example, be a human being), and in such cases subject locators would be totally inappropriate.

We are therefore faced with the following choice:

1. EITHER define a conditional mapping rule ("use a subject locator when the value of **identifier** is a locator which directly identifies the resource, and use subject identifier when it is a locator which indirectly identifies the resource"). Such a rule would be complicated to explain, subject to incorrect usage, and not easily amenable to automated processing.
2. OR define a simple rule ("use an occurrence"), which does not take full advantage of the semantics available in the Topic Maps model, but can be expected to be applied with great consistency.

In our opinion, consistency is the more important factor and so the current proposal mandates the second alternative.

3.6 Conclusion and Proposed Synthesis

Based on the preceding discussion of issues raised by earlier work, the approach to expressing DC metadata in Topic Maps proposed in this paper is as follows:

1. Regard Dublin Core resources as subjects.
2. Use statements to assign DC properties.
3. For each property, specify which kind of statement to use.
4. Recommend the use of associations for all elements *except* —
    * elements that are naming properties and thus more naturally represented as topic names (e.g. **title**); and
    * elements whose values consist of descriptive text or structured data types and are thus more naturally represented as occurrences (e.g. **description, date**).
    * For each property, specify which kind of statement to use.
5. Reuse existing identifiers.

The next section contains a complete set of declarations for all 85 Dublin Core terms in all four term sets, and also declarations for role types.

4 Declarations and Examples

4.1 Role Types

```
%prefix  dctm  http://psi.topicmaps.org/iso29111/

resource dc:resource - "Resource" .
value    dc:value    - "Value" .
```

4.2 DCMES

```
%prefix  dc  http://purl.org/dc/elements/1.1/

# name types
title dc:title - "Title"

# occurrence types
date dc:date - "Date"
description dc:description - "Description"
identifier dc:identifier - "Identifier"
rights dc:rights - "Rights"

# association types
contributor dc:contributor - "Contributor"
coverage dc:coverage - "Coverage"
creator dc:creator - "Creator"
format dc:format - "Format"
language dc:language - "Language"
publisher dc:publisher - "Publisher"
relation dc:relation - "Relation"
source dc:source - "Source"
subject dc:subject - "Subject"
type dc:type - "Type"
```

4.3 Elements and Element Refinements

The term set *Other Elements and Element Refinements* defines additional properties, many of which are refinements of the core 15 elements of the DCMES. Element refinements are represented in the same way as the element that they refine. The remainder are represented according to the principle given in the fourth bullet of section 3.6, above:

```
%prefix  dct  http://purl.org/dc/terms/

# name types
alternative dct:alternative - "Alternative"

# occurrence types
abstract dct:abstract - "Abstract"
accessRights dct:accessRights - "Access Rights"
available dct:available - "Available"
bibliographicCitation dct:bibliographicCitation - "Bibliographic
Citation"
created dct:created - "Created"
dateAccepted dct:dateAccepted - "Date Accepted"
dateCopyrighted dct:dateCopyrighted - "Date Copyrighted"
dateSubmitted dct:dateSubmitted - "Date Submitted"
educationLevel dct:educationLevel - "Audience Education Level"
extent dct:extent - "Extent"
instructionalMethod dct:instructionalMethod - "Instructional Method"
issued dct:issued - "Issued"
license dct:license - "License"
medium dct:medium - "Medium"
modified dct:modified - "Modified"
provenance dct:provenance - "Provenance"
tableOfContents dct:tableOfContents - "Table Of Contents"
valid dct:valid - "Valid"

# association types
accrualMethod dct:accrualMethod - "Accrual Method"
accrualPeriodicity dct:accrualPeriodicity - "Accrual Periodicity"
accrualPolicy dct:accrualPolicy - "Accrual Policy"
audience dct:audience - "Audience"
conformsTo dct:conformsTo - "Conforms To"
hasFormat dct:hasFormat - "Has Format"
hasPart dct:hasPart - "Has Part"
hasVersion dct:hasVersion - "Has Version"
isFormatOf dct:isFormatOf - "Is Format Of"
isPartOf dct:isPartOf - "Is Part Of"
isReferencedBy dct:isReferencedBy - "Is Referenced By"
isReplacedBy dct:isReplacedBy - "Is Replaced By"
isRequiredBy dct:isRequiredBy - "Is Required By"
isVersionOf dct:isVersionOf - "Is Version Of"
mediator dct:mediator - "Mediator"
references dct:references - "References"
replaces dct:replaces - "Replaces"
requires dct:requires - "Requires"
rightsHolder dct:rightsHolder - "Rights Holder"
spatial dct:spatial - "Spatial"
temporal dct:temporal - "Temporal"
```

4.4 Encoding Schemes

Encoding schemes (including DCMI Type) are always represented as topics.

```
%prefix dct http://purl.org/dc/terms/

Box dct:Box - "DCMI Box"

Dctype dct:Dctype - "DCMI Type Vocabulary"
DDC dct:DDC - "DDC"
IMT dct:IMT - "IMT"
ISO3166 dct:ISO3166 - "ISO 3166"
ISO639-2 dct:ISO639-2 - "ISO 639-2"
LCC dct:LCC - "LCC"
LCSH dct:LCSH - "LCSH"
MESH dct:MESH - "MeSH"
NLM dct:NLM - "NLM"
Period dct:Period - "DCMI Period"
Point dct:Point - "DCMI Point"
```

```
RFC1766 dct:RFC1766 - "RFC 1766"
RFC3066 dct:RFC3066 - "RFC 3066"
TGN dct:TGN - "TGN"
UDC dct:UDC - "UDC"
URI dct:URI - "URI"
W3CDTF dct:W3CDTF - "W3C-DTF"
```

When an encoding scheme is used to annotate the value of a property, the corresponding statement should be scoped accordingly, thus:

```
dc:subject ( dctm:resource : topicmap, dctm:value : Opera )
   @dct:LCSH
```

This association represents the dc:subject property whose value (represented by the topic with the ID Opera) is a term from the Library of Congress Subject Headings encoding scheme.

4.5 DCMI Type Vocabulary

The terms in the DCMI Type Vocabulary are represented as topics that play the role of 'value' in associations of type **dc:type**:

```
%prefix dctype  http://purl.org/dc/dctype/

Collection dctype:Collection - "Collection"
Dataset dctype:Dataset - "Dataset"
Event dctype:Event - "Event"
Image dctype:Image - "Image"
InteractiveResource dctype:InteractiveResource - "Interactive
   Resource"
MovingImage dctype:MovingImage - "Moving Image"
PhysicalObject dctype:PhysicalObject - "Physical Object"
Service dctype:Service - "Service"
Software dctype:Software - "Software"
Sound dctype:Sound - "Sound"
StillImage dctype:StillImage - "Still Image"
Text dctype:Text - "Text"
```

4.6 Example

The following example shows how to apply the preceding declarations in order to assign metadata to a topic map. We assume they are collected in a separate file called dc-declarations.ctm which is included in our topic map. The topic map is reified and statements are made about the reifying topic using the Dublin Core vocabulary:

```
%include dc-declarations.ctm
%prefix  o http://psi.ontopedia.net/

# reify the topic map
~ topicmap

topicmap
  - title : "DCMI Metadata Terms"
  date : 2007-07-04
  description : "A Topic Maps representation of DCMI Metadata Terms"
  source : http://dublincore.org/documents/dcmi-terms/
  isReplacedBy : "Not applicable"

creator( resource : topicmap, value : o:Steve_Pepper )
contributor(resource : topicmap, value : o:Patrick_Durusau )
contributor(resource : topicmap, value : o:Sam_Oh )
format(resource : topicmap, value : o:LTM )
```

5   The Subject-Centric Advantage

Having clarified the basic approach and defined the necessary vocabulary we now return to the issue raised in sections 3.1 and 3.2: the extent to which occurrences should be used rather than (or in addition to) associations. The following two examples demonstrate the two approaches:[8]

2-Legs Example (Not Recommended)

```
*e21 isa: entry
  - title: "Dublin Core in Topic Maps"
    date: 2007-08-07
    subject: "Subject-centric computing"
    subject: "Dublin Core"
    creator: "Steve Pepper"
    contributor: "Dmitry Bogachev"
    description: "Topic Maps and Dublin Core complement each other"
```

4-Legs Example (Recommended)

```
*e41 isa: entry
  - title: "About Dublin Core in Topic Maps"
    date: 2007-08-07
    subject:  Subject-centric_computing
    subject:  Dublin_Core
    creator:  Steve_Pepper
    contributor:  Dmitry_Bogachev
    description: "Topic Maps and Dublin Core complement each other"
```

The difference between these two examples may seem subtle but it can be extremely significant. In the 2-legs example, the values of the subject, creator and contributor properties are strings; in the 4-legs example, they are topics. For a single resource description viewed in isolation, the latter provides little extra value and involves some overhead. But for a description set consisting of multiple descriptions, the advantages of the subject-centric approach soon become apparent. The subjects "Subject-centric computing", "Dublin Core", "Steve Pepper" and "Dmitry Bogachev" represented by the topics in the 4-legs example become additional points of collocation and thus improve findability. Each of them can potentially be associated with any number of other subjects in multiple ways and thus new navigation paths become available. And each of these subjects offers a new potential for linking this topic map with information in other, related topic maps that may not necessarily represent Dublin Core metadata.

6   The DCMI Topic Map

In order to support the analysis performed in this paper, an easily navigable, richly hyperlinked, and queryable representation of the document *DCMI Metadata Terms* [3] was generated – as a topic map [13]. It includes the following:

- topics for each metadata term
- the properties of each term and its recommended form of representation in Topic Maps

---

[8] The code examples have been simplified for reasons of clarity. In particular, template declarations are assumed but not shown in the 4-legs example.

- associations linking each term to the set to which it belongs.
- associations linking elements to their refinements
- associations linking encoding schemes to the properties that they qualify

7 Conclusions

This paper has discussed previous work and the issues involved in representing Dublin Core metadata in Topic Maps. It has proposed a unified approach suitable for standardization by ISO and pointed out some of the advantages of the subject-centric approach.

References

[1] Dublin Core Metadata Element Set, Version 1.1. DCMI Recommendation (December 18, 2006), http://www.dublincore.org/documents/dces/

[2] ISO 15836-2003. Information and documentation — The Dublin Core metadata element set. ISO, Geneva (2003)

[3] DCMI Metadata Terms. DCMI Recommendation (December 18, 2006), http://www.dublincore.org/documents/dcmi-terms/

[4] Dublin Core Abstract Model. DCMI Recommendation (June 6, 2007), http://www.dublincore.org/documents/abstract-model/

[5] Expressing Dublin Core in HTML/XHTML meta and link elements. DCMI Recommendation (November 11, 2003), http://dublincore.org/documents/dcq-html/

[6] Guidelines for implementing Dublin Core in XML. DCMI Recommendation (April 2, 2003), http://dublincore.org/documents/dc-xml-guidelines/

[7] Expressing Simple Dublin Core in RDF/XML. DCMI Recommendation (July 31, 2002), http://dublincore.org/documents/dcmes-xml/

[8] Pepper, S.: NP for Technical Report - Information Technology - Topic Maps - Expressing Dublin Core Metadata using Topic Maps (2006), ISO/IEC JTC 1/SC 34 N0758 http://www.jtc1sc34.org/repository/0758.htm

[9] Algermissen, J.: A Processing Model for HTML using Dublin Core (November 19, 2001), http://www.topicmapping.com/psi/dc/dc.html

[10] Heuer, L., Hopmans, G., Oh, S., Pepper, S.: ISO/IEC CD, 13250-6 Information Technology - Topic Maps - Compact Syntax (CTM). ISO/IEC JTC 1/SC 34 N0905 (September 9, 2007), http://www.jtc1sc34.org/repository/0905.htm

[11] Pepper, S.: LTM example for Dublin Core metadata. v 1.4 dc-example.ltm (September 9, 2003)

[12] Maicher, L.: Dublin Core Metadata Terms. Topic Map (2006), http://www.informatik.uni-leipzig.de/~maicher/topicmaps/DCMT.ltm

[13] Pepper, S.: DCMI Metadata Terms Topic Map., http://www.ontopedia.net/omnigator/models/topicmap_nontopoly.jsp?tm=DublinCore.ltm

Appendix A

```
/* LTM example for Dublin Core metadata
   -----------------------------------
   $Id: dc-example.ltm,v 1.4 2003/09/18 12:55:46 pepper Exp $ */

#INCLUDE "dc.ltmm"
#TOPICMAP this-tm
[tm-topic = "LTM Dublin Core Example" @"#this-tm"]

/* DC properties modelled as occurrences: */
{tm-topic, Description, [[This topic map provides an example in LTM
of how to add Dublin Core metadata to a topic map.]]}
{tm-topic, Date, [[$Date: 2003/09/18 12:55:46 $]]}
{tm-topic, Rights, [[Copyright (c) 2003, Ontopia. This topic map may
be freely distributed in its current form provided this copyright
notice is left intact. Portions of this topic map may be copied,
modified, and reused without attribution for the purpose of adding
metadata to other topic maps.]]}

/* DC properties modelled as associations: */
Creator(tm-topic : resource, pepper : value)
Subject(tm-topic : resource, dc : value)
Subject(tm-topic : resource, metadata : value)
Subject(tm-topic : resource, topicmaps : value)
Publisher(tm-topic : resource, ontopia : value)
Contributor(tm-topic : resource, larsga : value)
Type(tm-topic : resource, topicmap-type : value)
Format(tm-topic : resource, LTM-format : value)
Source(tm-topic : resource, na : value)
Language(tm-topic : resource, en : value)
Relation(tm-topic : resource, na : value)
Coverage(tm-topic : resource, na : value)

/* Topic used for non-applicable DC properties */
[na = "N/A"]

/* Topics used as values of real metadata for this topic map */
[en       = "English"
   @"http://www.topicmaps.org/xtm/1.0/language.xtm#en"]
[ontopia  = "Ontopia" @"http://psi.ontopia.net/ontopia/#ontopia"]
[pepper   = "Steve Pepper"
@"http://psi.ontopia.net/ontopia/#pepper"]
[larsga   = "Lars Marius Garshol"
@"http://psi.ontopia.net/ontopia/#larsga"]
[dc       = "Dublin Core" @"http://psi.ontopia.net/metadata/#dc"]
[metadata = "Metadata" @"http://psi.ontopia.net/metadata/#metadata"]
[topicmaps = "Topic Maps"
   @"http://psi.ontopia.net/iso13250/#topicmaps"]
```

Mapping between the Dublin Core Abstract Model DCAM and the TMDM

Lutz Maicher

University of Leipzig, Johannisgasse 26, 04103 Leipzig, Germany
maicher@informatik.uni-leipzig.de

Abstract. Dublin Core is a widely used metadata vocabulary. There is ongoing standardisation work for the usage of DC-vocabularies in Topic Maps. Although the DC-vocabularies are already used in RDF, XML and HTML, they are defined independently of any particular encoding syntax. The DCMI Abstract Model (DCAM) is the metamodel for the DC vocabulary. It should facilitate the development of better mappings and interoperability. In a first step, this paper describes the relationships between the DCAM and the Topic Maps metamodel (TMDM). Afterwards, the directed DCAM→TMDM and TMDM→DCAM mappings are defined. The DCAM→TMDM mapping, combined with a serialisation specification for a Topic Maps notation, is an encoding guideline for this Topic Maps notation. For the ongoing standardisation of the usage of DC-vocabularies we propose a two layers approach. The first is the mapping defined here, which assures the interoperability between DC-metadata in Topic Maps and documented in other representation formats. The second layer provides authoring guidelines for Topic Maps authors describing DC-metadata. Strictly adhering to these authoring guidelines will assure that the created topic maps can be mapped to the DCAM and, much more important, become always be mergeable (irrespective of the DCAM in the background).

1 Introduction

Due to the subject-centricness of metadata, their representation in Topic Maps is obvious.[1] One of the most widespread metadata vocabularies is developed by the Dublin Core Initiative [1] (DCMI). Recently the standardisation of the *usage* of the DC-vocabulary in Topic Maps is urged on (see [7], [Pe08]).[2] This paper introduces a formal mapping between the metamodels of Topic Maps and the DC-vocabulary. We argue that the standardisation of the usage of DC terms in Topic Maps will be much more accurate, concrete and convenient on top of the mapping defined here.

Generally, the vocabulary specified by the DCMI is intended to be independent from the used representation methods, like HTML, RDF, or Topic Maps. This approach should assure the interoperability of metadata descriptions. For supporting this universality the DCMI has developed a metamodel for the DC-vocabulary, the Dublin Core Abstract Model (DCAM) [PNN+07].

---

[1] In [Pe08] Pepper illustrates this premise very well by comparing the power of so called 2-legs and 4-legs examples.
[2] Other related work is described in [Pe08] in the section "Reviewing of previous work".

L. Maicher and L.M. Garshol (Eds.): TMRA 2007, LNAI 4999, pp. 198–213, 2008.

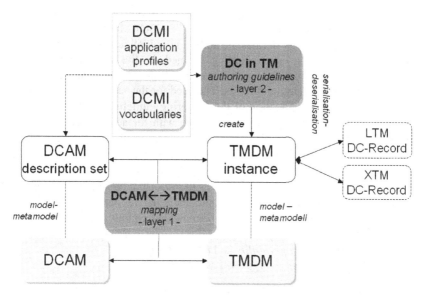

Fig. 1. Standardising the usage of DC-vocabularies in Topic Maps

As figure 1 illustrates, for the usage of the DC vocabulary (and all of its application specific extensions, the *application profiles*) in Topic Maps the relationship between the TMDM and the DCAM is important. According to the DCAM, all metadata about a *resource* is represented as a *description*. A bunch of related descriptions is called *description set*. Using standardised *encoding guidelines* these description sets can be serialised to RDF, XML, HTML or Topic Maps.

To publish a description set as topic map, it must be transformed into a valid TMDM instance. This transformation relies on a DCAM→TMDM mapping as defined in this paper. Afterwards, the produced TMDM instance can be serialised using a Topic Maps notation, like XTM, LTM or CTM. As figure 1 illustrates, each bundle of a DCAM→TMDM mapping and a serialization specifications for a Topic Maps notation is an *encoding guideline* for this Topic Maps notation.

But for empowering the full interoperability, the mapping must be both directed. In this case, even a topic map representing DC-metadata can be transformed into a description set. Unfortunately, the TMDM and the DCAM have a different terminological complexity. In example the TMDM provides typed and scoped names, which is completely unknown in the DCAM. Due to these differences, an isomorphic mapping between TMDM and DCAM is not possible. Therefore this paper introduces two directed mappings: DCAM→TMDM and TMDM→DCAM.

The recent standardisation efforts [7] are about the **usage** of DC-terms in Topic Maps. It is obvious, that when authoring a topic map these terms should be used in a way that always valid description sets can be produced by applying TMDM→DCAM. Only in this case, the *interoperability* to other representation formats like RDF or HTML as intended by the DCMI can be assured.

Summarised, when standardising the usage of DC-vocabularies we foresee the need of a two layers approach. The *first layer* is a DCAM⟵⟶TMDM mapping as defined here, the *second layer* defines the authoring guidelines for creating topic maps representing DC-metadata. The mapping defined in the first layer assures the interoperability of DC metadata expressed in Topic Maps and such metadata expressed in other representation formats. The need for the second layer is twofold.

On the one hand, the application of the authoring guidelines will assure that such created topic maps will always be interoperable with metadata represented in other representation formats. On the other hand, the application of the authoring guidelines will assures the *mergeability* of the created topic maps (irrespective of the DCAM in the background). This mergeability is due to the fact, that the authoring guidelines will exactly define how an observation, i.e. the information about the creator of a resource, has to be documented in a topic map. If all authors adhere to these guidelines, the same observations will be documented equally, with the result of the mergeability of the created topic maps [MB06]. The approach of this paper will assure both in parallel, *interoperability* and *mergeability*.

The reminder of this paper is organised as follows:

- *DCMI Meta Terms and the DCMT-Topic Map* provide a short overview about the terms defined by the DCMI, including a reference to the DCMT-topic map,
- *Dublin Core Abstract Model* provide a deep introduction into the three layers of the DCAM, including a comprehensive investigation into the relationships to the TMDM,
- *The identity crisis in the DCAM* discusses the differences in the identity disclosure mechanisms in DCAM and TMDM, including the implications for the defined mappings,
- *DCAM→TMDM* and *TMDM→DCAM* properly define the mappings, followed by a short *Example*, and
- *Towards authoring guidelines for DC in Topic* Maps describes in short detail requirements for the foreseen authoring guidelines as second layer of the standardisation.

2   DCMI Meta Terms and the DCMT-Topic Map

The DC-vocabulary is separated into the following five categories. All terms are not bound to any syntax, but their intended usage is defined in [DCMI]:[3]

- *Elements[4]:* contributor, coverage, creator, date, description, format, identifier, language, publisher, relation, rights, source, subject, title, type

---

[3] All terms labelled with (*) have the status „conforming"[6]. The necessity of these terms is proved by a specific user community. These terms are conforming to the DCAM, but do not belong to the core vocabulary.

[4] The set *elements* are the 15 basic terms of the DC vocabulary. These terms are additionally standardised as ISO 15836:2003 [ISO15836]. According to [DCMI] the identifier of an element is the namespace *http://purl.org/dc/elements/1.1/* suffixed by the according term. In example, the subject identifier for *coverage* is *http://purl.org/dc/elements/1.1/coverage*

- *Other Elements[5]:* accrualMethode[*], accrualPeriodicity[*], accrualPolicy[*], audience, instructionalMethod[*], provenance[*], rightsHolder[*]
- *Element Refinements[6]* abstract, accessRights[*], alternative , available, bibliographicCitation[*], conformsTo, created, dateAccepted[*], date Copyrighted[*], dateSubmitted[*], educationLevel[*], extent, hasFormat, hasPart, hasVersion, isFormatOf, isPartOf, isReferencedBy, isReplacedBy, isRequiredBy, isVersionOf, issued, license, mediator, medium, modified, references, replaces, requires, spatial, tableOfContents, temporal, valid
- *DCMI Type Vocabulary[7]:* Collection, Dataset, Event, Image, InteractiveResource, MovingImage, PhysicalObject, Service, Software, Sound, StillImage, Text
- *Vocabulary and Encoding Schemes[8]:* Box, DCMIType, DDC, IMT, ISO3166, ISO639-2, LCC, LCSH, MESH, Period, Point, RFC1766, RFC3066, TGN, UDC, URI, W3CDTF

All information about the DC-vocabulary given in [DCMI] is published in the "Dublin Core Metadata Terms"-topic map (DCMT-topic map) [Ma07].[9] Due to the DCMT-topic map authoring metadata in a topic map becomes more convenient, because it becomes always sufficient to only refer to any used term of the DC-vocabulary using the defined subject identifiers. Through merging in or requesting the DCMT-topic map (i.e. with TMRAP [Ga06]) all information about the terms, like labels, dependencies or definitions, can easily be acquired when needed.

3 Dublin Core Abstract Model (DCAM)

The Dublin Core Abstract Model (DCAM) [PNN[+]07] is the metamodel for the DC vocabulary. The DCAM has three layers: the DCMI resource model, the DCMI Description set model, and the DCMI vocabulary model. In the following these models will be introduced and the relationships to the TMDM as metamodel for Topic Maps will be investigated in detail.

---

[5] *Other elements* are elements, which do not belong to the 15 core terms of the DC vocabulary. According to [DCMI] the identifier of an other element is the namespace *http://purl.org/dc/terms/* suffixed by the according term.

[6] *Element Refinements* are terms which specify Elements in more detail. According to [DCMI] the identifier of an element refinement is the namespace *http://purl.org/dc/terms/* suffixed by the according term.

[7] The *DCMI type vocabulary* is a generic, domain independent vocabulary for typing resources. According to [3] the identifier of a type term is the namespace *http://purl.org/dc/dcmitype/* suffixed by the according term.

[8] At the moment, the DCMT-topic map does not contain any information about vocabulary and enconding schemes. This will change in future.

[9] Besides the officially defined identifiers in [DCMI] the DCMT-topic map additionally defines identifiers for each term of the DC-vocabulary in *one* namespace *http://psi.semports.org/dc/*. The reason for introducing these additional identifiers is convenience for the Topic Maps authors, because all terms are within one namespace. This avoids numerous confusions for the authors. Referring Occam's razor Pepper argues, that such a duplication should be avoided. But as shown in [Ma07a, Ma07c], the synonymous usage of subject identifiers is unproblematic, if the synonymy is disclosed as it is done by the DCMT-topic map.

3.1 DCMI Resource Model

The *DCMI resource model* specifies how the "real" world is composed when looking through DCAM glasses. According to the DCMI resource model, the whole world is a set or *resources*. A resource is "anything that has identity". The DCMI resource interpretation is equivalent to resource definition in RDF [RDF]. Pursuing this chain and taking the RDF←→TMDM [PPG⁺06] discussion into account it becomes obvious, that a *resource* in DCAM is the same as a *subject* in Topic Maps (see [DN07]).

According to the resource model, a resource is composed of *property/value pairs*. A *property* is a specific aspect, characteristic, attribute, or relation used to describe the resources. To each property a *value* is assigned, which is by definition a resource. In Topic Maps, not any assumptions about the composition of the subjects in the "real world" are defined anywhere.

Each resource becomes a *described resource* when a proxy is created and information is documented about it. The DCMI description set model in the next section defines the structure of these proxies, which are called descriptions.

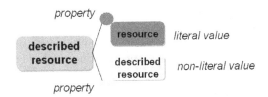

Fig. 2. The DCMI Resource Model

Furthermore, the DCMI resource model separates between *literal values* and *non-literal values* of properties. The DCAM describes this difference as follows: "Each *non-literal value* may be the *described resource* in a separate *description* within the same description set – for example, a separate *description* may provide metadata about the person that is the creator of the *described resource*. A *literal value* can not be the *described resource* in a separate *description*." [PNN⁺07] Summarised, the resource which is a *non-literal value* is represented by a proxy in the models and the resource which is a *literal value* is only represented as a literal in the models. As Figure 2 illustrates, only the resource of a non-literal value becomes a described resource by its own.

3.2 DCMI Description Set Model

The *DCMI description set model* specifies how information about resources – which are sets of property/value pairs – will be represented in the description sets. Generally, the description set model defines the modelling constructs of the DCAM. To most extend, the DCAM←→TMDM mapping is a mapping between the DCMI description model and the TMDM.

From the DCAM perspective, the proxy of a described resource is a *description*. The counterpart in the TMDM is a topic, which is a proxy for a subject. Associated

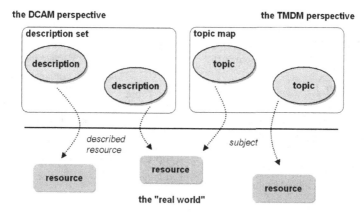

Fig. 3. Relationship between descriptions and topics

descriptions can be combined into *description sets*. The counterpart in the TMDM are topic maps, seen as TMDM instances. Figure 3 summarises these similarities. From the DCAM perspective a resource of the "real world" plays the role of "described resource" for its proxy (which is a *description* there). From the TMDM perspective the same resource plays the role "subject" for its proxy (which is a *topic* there).

According to the DCMI description set model, each description identifies the described resource by zero or **one** *described resource URI*. In contrast, in the TMDM the subject is described by a set of subject identifiers or subject locators. As discussed in the section "Identity crisis in the DCAM" this asymmetry avoids an isomorphic DCAM←→TMDM mapping.

The basic modelling constructs of descriptions are *statements*. Generally, each statement is a proxy for a property/value pair of a described resource. In the TMDM, topics are even used to make statements about subjects. In contrast to the very generic nature of any statements in the DCAM, the TMDM differs between topic names, variant names, occurrences, and associations. This even avoids an isomorphic DCAM←→TMDM mapping.

As shown in Figure 4, each statement in the DCAM is a composition of a *property URI* and a *value surrogate*. These property URIs are the identifiers defined by the DC vocabulary (see above), like *http://purl.org/dc/elements/1.1/coverage* for the term *coverage*. According to the resource model *each* value is a resource, so the value surrogate is always a proxy for the value.

Following the separation of literal and non-literal values in the resource model, the description set model differentiates between *literal value surrogates* and *non-literal value surrogates*. Each literal value surrogate is one *value string* (which is defined in detail below). Mostly a non-literal value surrogate is a value URI. This URI is a reference[10] to the description which represents the non-literal value. Furthermore, an

---

[10] There is a subtle difference to the TMDM. In the DCAM the value URI is a reference. If there exists another description which uses this value URI as described resource URI, a valid reference is established. Otherwise, in the TMDM all values of properties which refer to other information items *are* these items, and not only references.

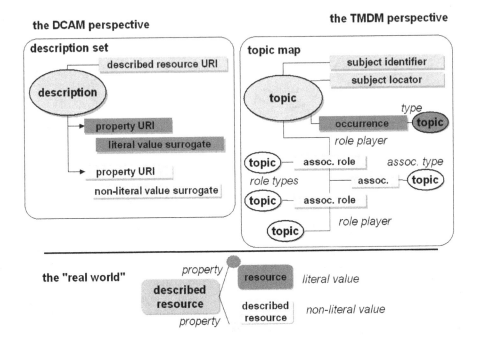

Fig. 4. DCAM and TMDM relationship

encoding scheme[11] *URI* can be assigned to a non-literal value surrogate. This URI is used to identify the vocabulary the used term (the value URI) is from. Alternatively to a value URI, the resource which is the non-literal value can also be described by a set of *value strings*.

From the TMDM perspective a property/value-pair having a literal value should be represented by an occurrence. Each occurrence is composed of a value (plus its datatype), which is the proxy for the literal value, and its type (which is another topic), which is the proxy for the property.

From the TMDM perspective a property/value-pair having a non-literal value should be represented by an association. Each association is composed by a set of roles and a type. The topic which is the type of the association is the proxy for the property. In the TMDM each role is composed by a type and a role player. In the property/value relationship between a resource and non-literal value, the topic for the resource plays one role and the topic for the non-literal value plays the opposite role. The description set model is agnostic to the role types to use, they will be defined in the DCAM→TMDM mapping. Figure 4 summarises the similarities of the DCMI description set model and the TMDM.

Generally, there exists a further distinction between the DCAM and the TMDM. The TMDM defines for all information items equality and merging rules. In

---

[11] "Vocabulary Encoding Schemes indicate that the value is a term from a controlled vocabulary, such as the value 'China – History' from the Library of Congress Subject Headings." [5].

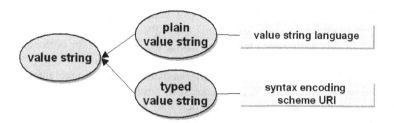

Fig. 5. The composition of a value string

consequence, two topic items representing the same subject (according to the defined rules) will be merged (according to the defined rules). This approach for using the identity of the proxies is not applied by the DCAM.

As last part the description set model defines value strings. A value string can be either a *plain value string* or a *typed value string*. A plain value string is intended to be human readable and may be tagged with the language used in this string. For tagging, the ISO language tags (like en-GB) should be used. A typed value string is tagged by a *syntax encoding scheme URI*. This URI identifies the syntax encoding scheme[12] under which the string should be interpreted.

A serialised description set is called *record*. There are defined encoding guidelines for diverse representation formats like RDF, XHTML and XML. The DCAM⟷TMDM mapping in the following bundled with the serialisation specification of a Topic Maps notation is an encoding guideline for this notation.

3.3 The DCMI Vocabulary Model

In a last step, the DCMI vocabulary model is defined, which is an abstract model of the vocabularies *used* in the descriptions. According to this model, each vocabulary is a set of terms, which can be properties, classes, syntax encoding schemes, or vocabulary encoding schemes. Between these terms sub-property, sub-classes and type-instance relationships can be defined. Because the description set model is used to represent the relationships between the terms, it is rather in the scope of the second layer of the standardisation - the authoring guidelines - and not in the focus of this paper.

4   The Identity Crisis in the DCAM

As highlighted by Pepper and Schwab [PS03], using URIs for identifying resources might result in an identity crisis. The reason for such a crisis is due to the non-existence of the distinction between addressable and non-addressable subjects (or resources). This problem is well known from RDF (the related discussion is sketched in [PS03]), where one URI can be used to identify the retrievable information resource itself (to make statements about this information resource, like metadata) or

---

[12] "Syntax Encoding Schemes indicate that the value is a string formatted in accordance with a formal notation, such as '2000-01-01' as the standard expression of a date." [5].

this URI can be used to identify the subject which is described by the retrievable information resource (to make statements about the subject which is represented by the content of this information resource). Using the same URI for both homonymous interpretations directly leads to confusions and merging errors.

The DCAM does not introduce a mechanism for the disambiguation of URI semantics, whereas the TMDM introduces such a mechanism through the separation of subject locators and subject identifiers. It is obvious that this has direct consequences for the DCAM$\leftrightarrow$TMDM mapping.

When mapping a description to a topic map, a decision must be done whether the *described resource URI* should be used as subject identifier or as subject locator in the created topic item. In some cases the usage of a subject identifier might be appropriate (i.e. when information about a person is represented by the description), in some cases the usage of a subject locator might be appropriate (i.e. when information about a retrievable information resource is represented by the description).

Obviously, this decision is application depended, it depends on the used vocabulary. For example the value of the property *dc:creator* is mostly a person, so the topic representing this non-literal value surrogate should use the described resource URI as subject identifier. For other terms, other rules might apply. For this reason, the DCAM$\leftrightarrow$TMDM mapping will not make any appointments about it, the definition of all such rules are delegated to the standardisation of the authoring guidelines. For example, in section "The identity element" of [Pe08] Pepper defines a specific rule for the authoring guideline specified there.

The inverse mapping direction from a topic item to a description yields additional problems. A topic item might have more than one *subject identifiers* or *subject locators*. (Due to the free mergeability of topic maps the authors will hardly be able to assure that a topic item has always only one sole identifier). For this reason, when mapping a topic item into a description, one of these identifiers might randomly be chosen to be used as described resource URI in the description. This might yield a loss of information.

To avoid such a deranging forfeiture of information, it should be recommended to document the URI which desired as described resource URI as occurrence typed by *dc:identifier* as it is proposed by Pepper [Pe08], too. But it should not be an error when such an occurrence is not assigned to a topic item. In this case, the mapping algorithm will chose one subject identifier or subject locator randomly.

5   The DCAM$\rightarrow$TMDM Mapping

The purpose of the DCAM$\rightarrow$TMDM mapping defined in the following is the transformation of a description set into a topic map. Each description D within a description set requires a topic item r in the topic map (which is seen as TMDM instance). This topic r will not be typed as "described resource proxy" because each topic should always be an agnostic representative of a subject, allowing any kinds of statements about it. The DC-relatedness of each statement is always derivable from the vocabulary used in these statements (see the TMDM$\rightarrow$DCAM section).

> **Note:** When stated in the following "create an [topic map construct] of type
> [URI]" means, that to the value of the property *c.[type]* of the topic

map construct a topic item *t* will be set, which has the URI as value of its property *t.[subject identifier]*.

When mapping a description D to a topic map, in the first step the topic item *r* representing the described resource must be created.

(a) **Create the resource topic.** Create a new topic item *r* which will be the proxy of the resource in the topic map.

(b) **Assign identity to the resource topic.** If assigned to D, the *described resource URI* will be set to *r.[subject locators]* (if this URI should be used as subject locator) or *r.[subject identifiers]* (if this URI should be used as subject identifier). According to the discussion in the previous section, the decision about the use of the *described resource URI* as subject locator or subject identifier is delegated to the authoring guidelines. As only constraint defined here, the decision must be compliant to the TMDM semantics.

(c) **Assign identity of the resource topic as occurrence.** Create an occurrence item *o* of type *dc:identifier*, assign the *described resource UR* to *o.[value]* and add *o* to *r.[occurrences]*.

In the next step each statement S of the description D must be mapped into the topic map. If a statement S has a non-literal value surrogate an association must be added to the topic map. In this case, the following has to be done:

(a) **Create the value topic.** Create a new topic item *v* which will be the proxy of the non-literal value in the topic map.

(b) **Assign identity to the value topic.** If available, assign the *value URI* of the non-literal value surrogate to *v.[subject locator]* (if this URI should be used as subject locator) or to *v.[subject identifier]* (if this URI should be used as subject identifier). The decision about the usage of the value URI as subject locator or subject identifier is not be specified here, but the decision must be compliant to the TMDM semantics.

(c) **Assign identity of the value topic as occurrence.** If *a value URI* is available, create an occurrence item *o* of type *dc:identifier*, assign the value URI of the non-literal value surrogate to *o.[value]* and add *o* to *v.[occurrences]*.

(d) **Assign vocabulary encoding scheme URI as occurrence.** If a *vocabulary encoding scheme URI* is available, create an occurrence item *o* of type *principles:vocabulary-encoding-scheme*, assign the vocabulary encoding scheme URI to *o.[value]* and add o to *v.[occurrences]*.

(e) **Assign value strings as occurrences.** If available, for each value string which is part of the non-literal value surrogate an occurrence item *o* of type *iso29111:valuestring* must be created according the guidelines for literal value surrogates below. Each o must be set to *v.[occurrences]*.

(f) **Create the typed association.** Create a topic item *at* and assign the *property URI* of the statement S to *at.[subject identifiers]*. Create an association item *a* in the topic map and assign the typing topic item *at* to *a.[type]*.

(g) **Create the resource role.** Create an association role item *ar1* of type *iso29111:resource*[13], assign *t* to *ar1.[player]* and assign *ar1* to *a.[roles]*.

---

[13] The PSI for the role resource as well as the PSI for the role type value are proposed by Pepper [Pe08].

(h) **Create the value role.** Create an association role item *ar2* of type *iso29111:value*, assign *v* to *ar2.[player]* and assign *ar2* to *a.[roles]*.

Note: The definition of the role types *iso29111:resource* and *iso29111:value* is necessary due to the directed nature of the statements in DC. For example, in an association of type *dc:creator* it is necessary to know which topic is the starting point (the created resource) and which topic is the endpoint (the creator of the resource). The authors of topic maps are completely free to create and use different role types such as *creator* or *publisher*. To be complaint to this document, these types must be subtypes (according to section 7 of [TMDM]) of *iso29111:resource* or *iso29111:value*.

If a statement S has a literal value surrogate a typed occurrence must be added to the topic map. In this case, the following has to be done:

(a) **Create and add the typed occurrence item.** Create a new occurrence item *o* which will be the proxy of the literal value in the topic map and add *o* to *r.[occurrences]*. The type of *o* is defined through the *property URI* of the Statement S.

(b) **Add value.** Add the *value string* of the Statement S to *o.[value]*.

(c) **Add syntax encoding scheme URI.** If available, add the *syntax encoding scheme URI* to *o.[datatype]*.

(d) **Add value string language as scope.** If a value string language is available, a topic item *os* must be created. The value of os.*[subject identifiers]* is set to the value string language as described below and *os* is set to *o.[scope]*.

Note: According to the DCAM the language should be indicated using a "ISO language tag". Such a language tag is a language abbreviation according ISO 639. For creating subject identifiers, the namespace *lang* should be suffixed by these acronyms.

Finally, two further points should be outlined.

(a) **Naming the resource topic.** For a better readability of the created topic maps, the value of a property which has naming characteristics, like *title*, can be *additionally* assigned as unscoped and untyped name of *r*. But the value of such a property *must* additionally be represented as defined above, its usage as topic name is only informative.

(b) **Typing the resource topic.** For a better readability of the created topic maps, the type of a resource can be *additionally* represented as type-instance relationship, where *r* is playing the role of an instance. But in general all typing properties *must* always be represented as defined above, its further representation in such a type-instance relationships is only informative.

6 TMDM→DCAM Mapping

The purpose of the TMDM→DCAM mapping defined in the following is the transformation of topic maps (which represent metadata using the DC vocabulary) into description sets. Due to the different terminological diversity of TMDM (which is multifaceted) and DCAM (which is more focused) such a transformation might always imply a loss of information (in most cases subject identifiers will be lost).

Furthermore, the mapping is error prone. Having the DCAM→TMDM mapping above, it will always be possible to create a valid topic map from any valid description set. But having a valid topic map, it is not possible to always create a valid description set out of this. For this reason, this mapping defines constraints which must be fulfilled by topic maps statements to assure that descriptions can be created out of them. If the constraints are not fulfilled, errors occur and statements of description can not be created.

The TMDM→DCAM mapping does only transform such information into the description set which is documented in the topic map using the DC vocabulary (or an approved application profile). All other information which is documented in the topic maps by using terms from other vocabularies (or terms of the DC vocabulary which are not used in the specified way) are *ignored* by the mapping defined here.

In a first step, for each topic item r, which represents a described resource according to the DCAM, a new description D must be created. Because topic items are not typed as "described resource proxy", eligibility is determined by the following rules:

(a) **Occurrence item.** The topic item r has at least one occurrence item o in its property *r.[occurrences]* which is typed by a term from the DC vocabulary (see above) or any approved application profile.

(b) **Role Player.** The topic item r plays at least in one association a role of type *iso2911:resource*. Furthermore this association must be typed by a term from the DC vocabulary (see above) or any approved application profile.

For each eligible topic item r a description D in the description set will be created. In a first step the identity must be assigned to the description:

(a) **Using the dc:identifier occurrence.** If available, the value of an occurrence item o in *r.[occurrences]* which is of type *dc:identifier* will become the *described resource URI* of the description D.

(b) **Using the subject identifiers or locators.** If such an occurrence is not available, one value from *r.[subject identifiers]* or *r.[subject locators]* can be used as *described resource URI* of the description D. If one of these values is a term from the DC vocabulary (or approved application profiles) this should be taken preferably, otherwise one value will be chosen randomly.

(c) **Using item identifiers.** If even subject identifiers or subject locators are not available, one randomly chosen value of *r.[item identifiers]* should be used as *described resource URI* of the description D.

In the next step all statements of D have to be created. The needed information is either documented in typed occurrences or in typed associations.

If an occurrence item o from *r.[occurrences]* is typed by a term from the DC vocabulary (or an approved application profile) a new statement S with a literal value surrogate will be added to D as follows:

(a) **Add property URI.** One subject identifier of the topic item ot in *o.[type]* which is element of a DC vocabulary (or an approved application profile) will be used as *property URI* in S. If such a subject identifier is not available, an error occurs and the statement S can not be created.

(b) **Add value string.** The value string of the statement S is the value of *o.[value]*.

(c) **Add syntax encoding scheme URI.** If the value of *o.[datatype]* is not *XMLSchema:string* the *syntax encoding scheme URI* of the value string in statement S is set to this value. In this case, the *literal value surrogate* is a *typed value string*. Otherwise, the literal value surrogate is a *plain value string*.

(d) **Add value string language.** If the literal value surrogate is a plain value string and *o.[scope]* contains a topic item which has a subject identifier in the namespace *lang*, the *value string language* of the value string in statement S is set to the part of this identifier after the namespace *lang* (i.e. en-GB).

If an association item *a* is typed by a term from the DC vocabulary (or an approved application profile) and *r* is playing the role of type *iso2911:resource* in this association (whereby the topic item *v* is playing the role of type *iso2911:value*) a new statement S with a non-literal value surrogate will be added to D as follows:

(a) **Add property URI.** One subject identifier of the topic item *at* in *a.[type]* which is element of a DC vocabulary (or an approved application profile) will be used as property URI in S. If such a subject identifier is not available, an error occurs and the statement S can not be created.

(b) **Add value URI by occurrence.** If *v* has an occurrence *o* typed by *dc:identifier*, the value of *o.[value]* is used as *value URI* of the statement S.

(c) **Add value URI by identifiers.** Otherwise, if *v* has values in the properties *v.[subject identifier]* or *v.[subject locators]*, the *value URI* of the statement S is set to one of these values. If one of these values is a term from the DC vocabulary (or approved application profiles) this should be taken preferably, otherwise one value will be chosen randomly. If none value is available, none value URI is assigned to the non-literal value surrogate of the statement S.

(d) **Add vocabulary encoding scheme URI.** If *v* has an occurrence item *o* which is typed by *principles:vocabulary-encoding-scheme* the *vocabulary encoding scheme URI* of the non-literal value surrogate is set to the value of *o.[value]*. If more then one occurrence items of this type are assigned to *v*, one of them has to be chosen randomly.

(e) **Add value strings.** If *v* has occurrence items *o* which are typed by *iso29111:valuestring* for each of these items a *value string* of the non-literal value surrogate is created according to the specification for literal value surrogates above.

(f) **Error checking.** If neither a *value URI*, a *vocabulary encoding scheme URI* nor a *value string* can be assigned to the statement, an error occurs and the statement S can not be created.

7 Example

In this section a short example illustrates the mapping between DC descriptions and Topic Maps. The DC description provides metadata about a book. It is composed of one statement with a literal value surrogate and two statements with non-literal value surrogates. (For saving space, the necessary descriptions of the creators of the book are not given in this example).

described resource URI	**doi:10.1007/11676904**
property URI	http://purl.org/dc/elements/1.1/type
value URI	http://purl.org/dc/dcmitype/Text
vocabulary enc. scheme URI	http://purl.org/dc/dcmitype/
property URI	http://purl.org/dc/elements/1.1/title
plain value string	Charting the Topic Maps Research and Applications Landscape
value string language	en-GB
property URI	http://purl.org/dc/elements/1.1/creator
value URI	mailto:maicher@informatik.uni-leipzig.de
property URI	http://purl.org/dc/elements/1.1/creator
value URI	mailto:jack.park@sri.com

Now the same information is presented as topic map, serialised in LTM 1.3 [8]. It should be underlined, that some information in a TMDM instance (like a datatype of an occurrence item) can not be represented using LTM. To get more information about the used terms from the DC-vocabularies the DCMT-topic map [Ma07b] can be merged in or requested.

```
#PREFIX dc @"http://purl.org/dc/elements/1.1/"
#PREFIX dctype @"http://purl.org/dc/dcmitype/"
#PREFIX lang @"http://www.topicmaps.org/xtm/1.0/language.xtm#"
#PREFIX iso29111 @"http://psi.topicmaps.org/iso29111/"
[id1 : dctype:Text = "Charting the Topic Maps ..." @"doi:10.1007/11676904"]
    {id1 , dc:title, [[Charting the Topic Maps Research ...]]} /lang:en-GB
    {id1 , dc:identifier, [[doi:10.1007/11676904]]}
dc:type(id1 : iso29111:resource, dctype:Text : iso29111:value)
[id2 @"mailto:maicher@informatik.uni-leipzig.de"]
    {id2 , dc:identifier, [[mailto:maicher@informatik.uni-leipzig.de]]}
[id3 @"mailto:jack.park@sri.com"]
    {id3 , dc:identifier , [[mailto:jack.park@sri.com]]}
dc:creator(id1 : iso29111:resource , id2 : iso29111:value)
dc:creator(id1 : iso29111:resource , id3 : iso29111:value)
```

8 Towards Authoring Guidelines for DC in Topic Maps

As already discussed in the introduction, for the standardisation of the DC/TM interoperability we foresee a two layers approach. The first layer defines the DCAM←→TMDM mapping, as it is realised in this paper. This mapping assures the *interoperability* of DC metadata expressed in Topic Maps and metadata expressed in other representation formats. The second layer is the definition of authoring guidelines for all terms of the DC-vocabularies for the creation of topic maps. The need for this layer is twofold. On the one hand, the application of these authoring guidelines will assure that such a created topic map will be interoperable with metadata represented in other representation formats. On the other hand, the application of the authoring guidelines will assure the *mergeability* of the created topic maps (irrespective of the DCAM in the background).

The standardised authoring guidelines should look like as follows:

First, it *must* be defined how a described resource and statements with literal-value and statements with non-literal values have to be represented when authoring a topic map. This standardisation must be strictly compatible to the TMDM→DCAM mapping defined here.

Second, for each term of the DC-vocabularies it *must* be decided, in which cases it should used as a property for a non-literal value and when it should be used as a property for a literal-value. No further specifications are mandatory for any term.

Third, guidelines for the representation of the described resources which are non-literal values *might* be defined. For example, best practice for choosing identifiers of persons, countries, dates, etc. can be defined. The more specific these specifications are, the better the mergeability of the resulting topic maps.

Fourth, it must be defined how relationships between terms of the DC-vocabulary (i.e. sub-property relationships, etc.) should be represented in Topic Maps to be compliant to the DCMI vocabulary model.

9 Conclusion and Further Work

This paper has introduced a comparison and a mapping of the metamodel of the Dublin Core metadata vocabulary, the Dublin Core Abstract Model, and the metamodel of Topic Maps, the Topic Maps Data Model. Due to the different terminological expressivity, an isomorphic mapping between DCAM and TMDM is not possible. As consequence, two directed mappings has been introduced here. Especially the TMDM→DCAM mapping might imply a loss of information.

The purpose of the defined mappings is the assurance of the interoperability between DC metadata expressed in Topic Maps and DC metadata expressed in other representation formats.

On top of these mappings the authoring guidelines for DC-vocabularies in Topic Maps should be standardised. Defining such modelling methodologies assures, that *(1)* the created topic maps are always interoperable with the DCAM and *(2)* all created topic map are mergeable (irrespective of the DCAM in the background).

On the short term we foresee two further work packages. *First*, the TMDM→DCAM mapping allows the specification of a *schema* which decides, whether a topic map represents DC-metadata correctly. Once TMCL, Topic Maps constraint language, is standardised, such a schema should be defined. Second, the specification of a DC metadata *filter* (view), which is a set of TMQL, the Topic Maps query language, queries, is very close to this approach. Applying these queries will extract exactly and only the DC metadata represented in a topic map.

References

[DCMI] DCMI Usage Board: DCMI Metadata Terms (2006),
   http://dublincore.org/documents/dcmi-terms/

[DN07] Durusau, P., Newcomb, S.: The Essentials of the Topic Maps Reference Model (TMRM). In: Maicher, L., Sigel, A., Garshol, L.M. (eds.) TMRA 2006. LNCS (LNAI), vol. 4438, pp. 152–160. Springer, Heidelberg (2007)

[Ga06] Garshol, L.M.: TMRAP – Topic Maps Remote Access Protocol. In: Maicher, L., Park, J. (eds.) TMRA 2005. LNCS (LNAI), vol. 3873, pp. 53–68. Springer, Heidelberg (2006)

[Ma07a] Maicher, L.: The impact of semantic handshakes. In: Maicher, L., Sigel, A., Garshol, L.M. (eds.) TMRA 2006. LNCS (LNAI), vol. 4438, pp. 140–151. Springer, Heidelberg (2007)

[Ma07b] Maicher, L.: Dublin Core Meta Terms. Topic map, http://informatik.uni-leipzig.de/~maicher/topicmaps/DCMT.ltm

[Ma07c] Maicher, L.: Autonome Topic Maps. Doctoral thesis, University of Leipzig (2007), http://www.informatik.uni-leipzig.de/~maicher/publications/DISS_LutzMaicher_german.pdf

[MB06] Maicher, L., Böttcher, M.: Closing the Semantic Gap in Topic Maps and OWL Ontologies with Modelling Workflow Patterns. Journal of Universal Computer Science 12(Special Issue I-Know 2006), 261–269 (2006)

[Pe08] Pepper, S.: Expressing Dublin Core in Topic Maps. In: Maicher, L., Garshol, L.M. (eds.) Scaling Topic Maps. LNCS (LNAI), vol. 4999, pp. 1–9. Springer, Heidelberg (2008)

[PNN+07] Powell, A., Nilsson, M., Naeve, A., et al.: DCMI Abstract Model,
   http://dublincore.org/documents/abstract-model/

[PPG+06] Pepper, S., Presutti, V., Garshol, L.M., et al. (eds.): Guidelines for RDF/Topic Maps Interoperability. W3C Editor's Draft.
   http://www.w3.org/2001/sw/BestPractices/RDFTM/guidelines-20060630.html

[PS03] Pepper, S., Schwab, S.: Curing the Web's Identity Crisis. Subject Indicators for RDF. In: Proceedings of XML, Europe (2003)

[RDF] Hayes, P. (ed.): RDF semantics. W3C recommendation,
   http://www.w3.org/TR/rdf-mt/

Webreferences

[1] http://dublincore.org/

[2] http://dublincore.org/documents/abstract-model/

[3] http://dublincore.org/documents/dcmi-type-vocabulary/

[4] http://dublincore.org/resources/expressions/

[5] http://dublincore.org/usage/documents/principles/

[6] http://dublincore.org/usage/documents/process/#conforming

[7] http://www.jtc1sc34.org/repository/0884.htm

[8] http://www.ontopia.net/download/ltm.html

Namespaces

iso29111   *http://psi.topicmaps.org/iso29111/*

lang   *http://www.topicmaps.org/xtm/1.0/language.xtm#*

principles *http://dublincore.org/usage/documents/principles/#*

XMLSchema *http://www.w3.org/2001/XMLSchema#*

On Path-Centric Navigation and Search Techniques for Personal Knowledge Stored in Topic Maps

Jens Heider and Julian Schütte

Fraunhofer Institute for Secure Information Technology, 64295 Darmstadt, Germany

Abstract. Leveraging interconnections across stored personal information is a novel concept to find desired information by context and structure. This paper introduces a combined navigation and search interface for homogeneous knowledge representations based on path and set calculation. It uses graph theory to provide a more intuitive way of supplying search criteria and retrieving related information.

1 Introduction

In many situations personal knowledge is stored in various locations and in different ways, requiring the user to use multiple applications to search for already known content. Even approaches providing search interfaces based on full-text search for multiple applications do not solve this sufficiently. They only shift the problem from finding the content to providing the right keyword contained in the content the user hopes to retrieve. Of course the office scenario of a knowledge worker handling e-mails, organizers, document management systems, and all other sorts of digital tools, trying to improve the daily work with information, is only one special domain to be covered.

This paper therefore introduces a search technique that leverages the redundancy of information contained in various data sources by autonomously building a homogeneous representation of it. Thus, useful connections between its metadata is created and used for a search based on graph theory. The approach presents one core aspect of the MIDMAY project[1] (Mobile Information Distribution Management and Access for You!) initially described in [1]. It builds the framework for mobile devices such as smartphones to evolve into an everyday tool for the secure distribution, management and access of information. The proof of concept implementation uses Topic Maps as the underlying technology for handling personal knowledge and its homogeneous merging [2].

The need for search techniques in Topic Maps is often addressed by keyword search. In applications aiming at technical users, also query languages or Query-by-Example approaches are proposed [3]. Instead, the approach to be presented here is based on marking topics remembered by the user to be related in some way to the information to be searched. The user follows the paths he has in his

---

[1] `http://www.project-midmay.de`

L. Maicher and L.M. Garshol (Eds.): TMRA 2007, LNAI 4999, pp. 214–225, 2008.
© Springer-Verlag Berlin Heidelberg 2008

mind to navigate to these topics in a quite natural way. Having marked two or more topics, the result set is then calculated using graph theory. Its application with Topic Maps is briefly described in Sect.2. Sect.3 shows the basic navigation methods inside the unified knowledge representation that are needed in order to be able to mark topics for the search process, which is described in Sect.4.

2 Preconditions

The topic maps used in the following are assumed to be generated from data sources containing information stored by the user. This is done with the help of software components, which we call extractors. They are built upon a framework, making it a simple task to provide access to new protocols of data sources, because only the mapping between the underlying information model to the access protocol has to be implemented. Each extractor is specialized this way and retrieves meta data from the data source to store it in a homogeneous topic map. Since each extractor is specialized on its data source, it can extract or create the meta data efficiently in the context of the extractors purpose. The resulting topic map structure and its transformation to a graph model is described in the following sections.

2.1 Topic Map Structure

The generated topic map contains resource references to data sources from where the actual information can be retrieved. The structure of the map therefore is designed to help the user find the desired references by preserving original structures he has given his data sources already. Additionally, the redundancy of data in multiple sources is used to interlink information in a way that enriches the representation. All information is then accessible via a single interface. In this context building a general representation with a homogeneous structure requires that

1. unified *Public Subject Indicators* (PSI) are used for a global typing schema across all extractors that is applied strictly to all topics,
2. associations also have a type to reflect their general semantic of a property, type or hierarchical relation for the global representation,
3. there is no directionality inherent in an association as defined by XTM specification [4] and
4. the topic map has to be made consistent, making each entry unique in the knowledge space of a user.

2.2 Graph Model

In contrast to a complete modeling of the entire Topic Maps concept as presented with the τ-model [5], we want to focus on the inherent structure of a topic map only. This has been done for directed acyclic graphs in [6], but in our application scenario we have to cover undirected cyclic graphs. The following model therefore represents our transformation from topic maps to a graph model that is useful for algorithms dealing with the concept of paths.

In our model the graph G, built by the topic map, is described by the pair (V, E). V is the finite set of vertices mapped to topics and E is a binary relation on V representing the undirected associations between the topics. Additionally, E explicitly contains the binary relations between topics and other Topic Maps' entities (e.g., topic types) to preserve this structural information. Reification is addressed by creating a new vertice in V, built out of the reified topic map construct (e.g., an association), and including its relations from all involved topics into E. Each edge $(v_i, v_j) \in E$ is given a constant configurable weight w_{ij} depending on the kind of connection and the search mode which will be described further in Sect.4.

A path of length k from topic t_a to topic t_b in this graph $G = (V, E)$ is then a sequence $\langle v_0, v_1, v_2, ..., v_k \rangle$ of vertices such that $t_a = v_0, t_b = v_k$ and $(v_{m-1}, v_m) \in E$ for $m = 1, 2, ..., k$. A path p_i connecting t_a and t_b is written as $(t_a \overset{p}{\leftrightarrow} t_b)_i$. The set P_{ab} of all n paths connecting the topics t_a and t_b is defined as $P_{ab} := \{\bigcup_{i=1}^{n} (t_a \overset{p}{\leftrightarrow} t_b)_i\}$. In this application field it can be assumed that the user is only interested in simple paths, so all paths in P_{ab} are defined to have distinct vertices to prevent cycles.

3   Navigation

As Topic Maps represent a network of topics interconnected by associations, it is obvious to use these paths for navigating through the node-centric network (see [7]). The interesting question is, which topic should be chosen as the starting node. Although no information about the search interest has been specified so far, the user has to be offered an entrance point to the topic map. Some approaches therefore use rich visualization to present several levels of detail for the representation. The user can enter it trough different dimensions, which supports the user in visually exploring the representation (see [8]). However, these approaches often overburden non-technical users and also require large displays.

Finding information stored in databases is often solved by query languages. A similar approach has also been developed for Topic Maps with the query language Tolog[2]. These query languages with their precise description capabilities of the desired result have much power. However, the inherent complexity also has some drawbacks for the non-technical user. He has to know the correct syntax and some attributes of the data structure to produce meaningful queries.

Instead, this approach is based on choosing topics known to be connected somehow to the desired information. These topics are selected during navigation inside the topic map. Therefore, the general concepts for navigation are described fisrt, before Sect.4 goes into detail about the search techniques and their background in graph theory.

---

[2] Implementation by Ontopia of the requirements stated by ISO Topic Maps query language TMQL standardization effort.

3.1   Points of Entrance

This section presents the methods we have integrated into MIDMAY to provide an easy way to enter the generated topic map for navigation. Each of them serves a slightly different purpose, depending on the kind of topic the user is looking for and the strategy he wants to follow.

Hierarchy Root. The most classic method to present an entry point is based on the hierarchy information contained in the topic map. The *Hierarchical Classification Pattern* and *Faceted Classification Pattern* proposed by Kal Ahmed [9] can be applied to unify a representation, as described in [2]. Therefore, all indexed data sources are represented in the topic map with their hierarchical structure. They share a common designated root that is presented to the user (see Fig.1). This way, the user is enabled to follow the paths to the leaves of the tree preserving the hierarchical structure. Of course, all other associations existing in parallel - not displayed in Fig.1 for the sake of clarity - can also be used to navigate within the topic map. This method is useful if the user remembers the hierarchical location of the original data source, which can now be accessed uniformly by using a single interface.

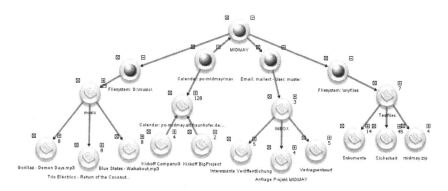

Fig. 1. Point of entrance using hierarchy root

Type List. This point of entrance leverages the typing information of all topics. Therefore, the user is presented with a list of all topic types contained in the topic map. By selecting one type, a list of all topics of that type (see Fig.2) is shown. In our scenario, this is useful for example to find documents of a certain file type regardless of its physical location in a data source. Additionally, this type list is useful to mark the complete topic type to be relevant for a search query instead of a single topic instance of the type.

Keyword Search. In other cases the user might remember a certain part of text contained in the metadata of the information he is looking for. Therefore, it is useful to provide an interface to enter keywords contained in the topic that represents the searched resource. To make it easier for the user, a personal

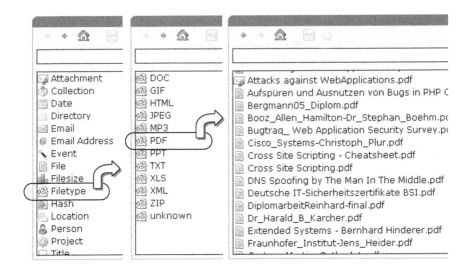

Fig. 2. Point of entrance using type lists

vocabulary is automatically generated from all topics by splitting the *Base Name Strings* into terms. This way, the user can browse through available keywords and the search interface can selectively provide suggestions for keywords while he types some letters of them. Especially on mobile devices with limited input interfaces, this improves the search process significantly. Void search queries are prevented and the input of the search criteria is sped up.

A second benefit of a search based on pre-calculated terms is the implicit generation of linguistic relations, which can assist the user find related information that is not explicitly linked by the topic map. Of course, these relations sometimes are not correct in the sense of the actual search, like in the case of homonyms or acronyms, but as the framework is designed for humans, it can be assumed that they are used to deal with those multiple meanings.

3.2   Navigation Aid

After the user has entered the topic map at the desired point, he moves from topic to topic by following the associations between them (see Fig.3(a)). In this node-centric interface (in MIDMAY called *Click and Cycle* because of its closed interaction cycle), a list of terms is presented at first. Choosing an entry leads to a list of topics containing the selected term. In this topic view, either a topic is marked for a topic operation such as using it inside a search query, or the topic is selected to see all of its associations. In this association view the user can select the desired association, which leads to the view of all related terms contained in associated topics. An additional redo phase is possible in all views in case a list would contain too many items to handle. In this case, the items will be presented in a staged view, clustering items alphabetically to provide a tradeoff between view levels and displayed items per level.

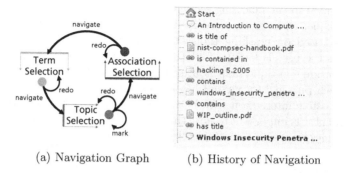

(a) Navigation Graph   (b) History of Navigation

Fig. 3. Navigating through the Topic Map by cycling the interface

To help the user keep track of the navigation we propose the integration of a path-centric approach. Its implications for the navigation will be shown in the next subsections. But it is also a key aspect for calculating search results as described in Sect. 4.

History Path. When navigating through personal knowledge it is often necessary to have the possibility to go back and forth between previously visited topics. To help the user find the desired topic he has already used for navigation, a history presents his path from the entrance point to the current position in the graph (see Fig. 3(b)). This path visualizes the relation of the intermediate topics and serves as a navigation shortcut to topics and associations. At the same time it can be displayed quite easily even on small displays. If more space is available on the display, the visualization can be enhanced with more sophisticated display techniques by showing adjacent topics in the surrounding of each contained node in the path as well.

Favorites. Topics used often can be marked as favorites. This is useful for example, if the topic represents a document used periodically, like the submitting of trade terms in reoccurring business processes. Choosing it from the list of favorites provides an instant access to the document, which can then be transmitted easily to any recipient without performing a search for it.

Depending on the data and its personal structuring it often also makes sense to use favorites for topics that are frequently used in search queries. This way selecting the required topics as input for the query can be reduced to only a few selections. Optionally it is also possible this way to store an actual selection of topics for a query to be easily reproducible for future requests.

4   Search

Besides navigation techniques, the inherent structure of a topic map offers other interesting aspects. These mainly arise from the linked information contained in the topic map. Every portion of information is associated with other instances.

This will be used in the following sections to provide an easy to use yet powerful search technique, without the need for the user to learn a query language. The approach is based on marking topics that are related in one way or another to the topics the user is looking for. To use a topic for a search, the described navigation inside the topic map is used to reach that desired topic. Then it is marked, which puts it in a virtual bag of accentuated topics. After this bag is filled, one possible interaction with it is using the search functionality.

At this point it is important to note that any topic can be used when specifying a query. Besides topics representing collected meta data, this also includes topics representing a type, such as *person*, *event* or *date*. As everything in a topic map is a topic, the user does not experience any restrictions nor must he perform different actions when formulating the search query, no matter what information he provides as input.

The next sections will describe the calculation of results based on marking topics. First the background for the used principle will be described, followed by the presentation of a simple algorithm we have used as proof of concept for this search technique. The shown sample search queries were performed with a MIDMAY prototype implementation presenting the results in its user interface.

4.1   Paths as Search Results

Let t_a be the first topic marked and t_b a second one. Because every topic has to have a topic type and the faced root topic connects all hierarchies, there is at least one path p with $t_a \overset{p}{\leftrightarrow} t_b$. For a simplified calculation we assume that paths with the smallest weight in P_{ab} are of the most interest. These paths represent the most direct relation in the context of the search, and paths with increasing weight are deemed to decrease in interest for the user.

Therefore every path $p \in P_{ab}$ connecting the topics t_a and t_b is sorted by its weight and displayed sequentially in that order starting with the lowest value. The paths are shown together with the descriptive name of the corresponding association to visualize the relation between them. The name is chosen by the role the topic plays to display the path readable from left to right or, for smaller displays, from top to bottom. Every topic and association of a path can then be used as an entry point for a subsequent navigation to their associations and topics respectively.

Path Calculation. A first simple approach only uses the topic map as an unweighted graph using $w_{ij} = 1$ for all edges $(v_i, v_j) \in E$. As a proof of concept an adapted *Bidirectional Search* is used that starts at t_a and t_b simultaneously. At both ends the connected vertices are inspected with a weighted *Breadth-first Search* in spreading waves, until the path with the lowest value is found. To produce the next path, the edge that connects both waves is removed from the search graph and a new path with lowest value is calculated. This significantly limits the amount of displayed paths, which is a desirable effect in this case, as the user only wants to see relations between the starting topics. In this case it is considered sufficient, if all topics in the possible paths are presented at

Table 1. Example weight values for path search modification

mode	$w(h)$	$w(p)$	$w(t)$	result
0	1.0	1.0	1.0	default case; shortest paths is calculated
1	0.7	2.0	2.0	hierarchical information in result path is desired
2	2.0	0.7	2.0	path should contain interconnection of properties
3	2.0	2.0	0.7	the interests in the relation of types is expressed

least one time in a displayed path, preventing the display of all other possible combinations of topics in paths computable in a dense connected graph. This way the user gets the chance to find any topic that is related to the marked topics t_a and t_b.

However, the topic map not only contains undirected edges represented by associations to be used for path calculation. The search can be further improved by taking the association types into account. These are assigned for all associations, assured by the extractor framework (see [2]). This way it is possible to distinguish for example between hierarchical relations and property relations during path calculation. By changing the weighting w_{ij} of the edges for an association type, the user is enabled to further specify his search interests.

An example for this is shown in Tab.1, which presents a weight assignment to the association types hierarchy (h), property (p) and type (t) to modify the search results. These values were chosen to deal with structures in the graph where multiple short paths connect two vertices. In the office scenario this is often the case because of a dense connection of topics (e.g., two topics t_a and t_b with associations to the same property topic t_c and an additional hierarchical relation between t_a and t_b). By setting the weight $w(x)$ of a relation type x to $\frac{2}{3} < w(x) < 1$ and the weights for the other types to 2, the paths through one or more of these typical three-cornered structures are selectable. Of course, this is not accurate in all and every case one can imagine, but these values have shown in many tests to work very well as an easy to use aid for the search process. By predefining these weights, the user only has to pick a descriptive mode to select the related weight assignment expressing his search interest.

Example: Leveraging Paths. The following simple example should explain how marking topics and calculating paths can be used as a search technique. The example is based on a topic map autonomously extracted by MIDMAY from a digital personal organizer, filesystem folders used for storing documents, and the personal e-mail folder. The user is looking for a document but can't remember its filename nor its author. However, the user recalls the town where he attended the conference and met the person who e-mailed the document. Provided that the personal organizer's calendar application contains the meeting along with the location and attendants' information, the extractors have already included this information into the topic map.

To use the path generation, the user navigates to the topics that build the endpoints of the query. In the example the location topic *Graz* and the filetype

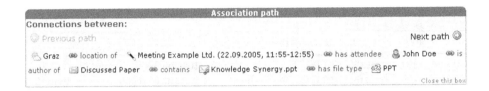

Fig. 4. Calculated shortest path example

topic *PPT* are marked, because the user is searching for a Microsoft Powerpoint presentation sent by someone he met at a meeting in the town Graz. The computed paths (see Fig.4) now show the resulting relations between these topics and the user can browse through the paths to find the one containing the topic referencing the attachment he is looking for. There is also the possibility to narrow the search in a second step by using the topics contained in the paths to navigate. After finding a path containing the name of the person who sent the attachment, the user can replace the location topic with this person topic. This excludes all other persons who also attended meetings in Graz.

Another strategy would be to mark topic *Graz* and topic *Person* - the type representing the concept of a person - to find all persons that are related to Graz. After selecting one person topic of the calculated paths, the search can be continued by marking other known aspects about the searched document.

4.2   Sets as Search Result

Section 4.1 assumes the selection of two topics to be used as query input. The next section deals with the case of a user wanting to provide more information in order to specify the desired result more precisely with three or more topics. The user will be presented with a set of topics that are considered related regarding the search parameters.

Calculation of Sets. To provide meaningful results in the case of multiple, marked topics, we decided to define the resulting sets based on the already defined shortest paths. First, all the shortest paths between all m marked topics in $S := \{s_1, ..., s_m\}$ are calculated to extract the relations between the topics. The resulting paths are then transformed to bit vectors. Therefore, all involved k vertices of all paths are indexed. These unique IDs are used as positions in all vectors. The vector B_{ab}^p indicates the presence or absence of vertices $\langle v_1, v_2, v_3, ..., v_k \rangle$ in path p between topic t_a and t_b. By using an OR operation defined as $B_{ab}^p \cup B_{ab}^{p'} := \{v_j | (v_j \in B_{ab}^p) \vee (v_j \in B_{ab}^{p'})\}$ on all bit vectors the resulting vector $B_{ab} = B_{ab}^1 \cup B_{ab}^2 \cup \cdots \cup B_{ab}^n$ represents the topics that are contained in all n shortest paths between t_a and t_b. This is performed for all pairs of the topics selected in S. The results are $\frac{m(m-1)}{2}$ bit vectors involved in the actual search. These bit vectors are then combined to a single one by applying a binary operation. In the default case this is an AND operation defined as $B_{ab}^{p} \cap B_{ab}^{p'} := \{v_j | (v_j \in B_{ab}^{p}) \wedge (v_j \in B_{ab}^{p'})\}$. The resulting vector $B_{res(S)} := \{\bigcap_{i=1,j=2}^{i=m-1,j=m} B_{s_i s_j} | s_i, s_j \in S, i < j\}$ represents the topics with a maximum of interestingness defined by the smallest weight of the paths to all of the marked topics. The topic selection view presents these topics as a result of the search query for further interaction.

Using this approach together with the weight assignment for the underlying path calculation presented in Tab.1, the user of the described scenario has the option to calculate sets based on

- mode 0, the default mode which only uses the structure of the topic map
- mode 1, to focus the search on topics equally connected by hierarchy
- mode 2, to focus the search on equal properties of marked topics
- mode 3, to focus the search on equal types of marked topics

Compared to query languages, this simple approach already produces valuable results and is easy to handle even for non-technical users. But of course the transformation into bit vectors was done to have the ability to expand the possibilities for search requests to other binary operations. In the first case the ordering of the marked topics does not matter. In an advanced mode it can be used to prioritize their meaning by specifying a binary operation between sets of marked topics. The operation is then used on the resulting bit vectors of the sets of marked topics. Other binary operations such as OR and XOR are defined similarly to the AND operation and produce according results. OR accumulates result topics, whereas XOR presents the difference between result sets.

Example: Marking Multiple Topics to Search. Imagine the situation of a user trying to find authors of presentations that are related to a certain project. His topic map is kept up to date automatically by MIDMAY and is composed by its extractors, indexing local document folders, his e-mail account and the company's LDAP[3] directory. The latter contains, besides many other useful information about the employees, their project memberships as well. Again the required information to answer the user's question is already stored in maintained data sources. By letting MIDMAY combine them in a personal knowledge map represented by a topic map, the user can use a single interface to mark topics that come to his mind when thinking of the desired result. He marks the topic *Person* and the topic *MIDMAY* that represents the project he is interested in (see fig.5). This would show all persons involved in the project MIDMAY. But by also marking the topic *PPT*, representing the type "PowerPoint presentations", the result is narrowed to persons related to this file type, too. The user can now navigate from this topic inside the topic map or browse for additional, related information such as e-mail addresses or telephone numbers. Additionally, the user can retrieve or distribute the information the topic stands for.

---

[3] Lightweight Directory Access Protocol, an application protocol for querying directory services.

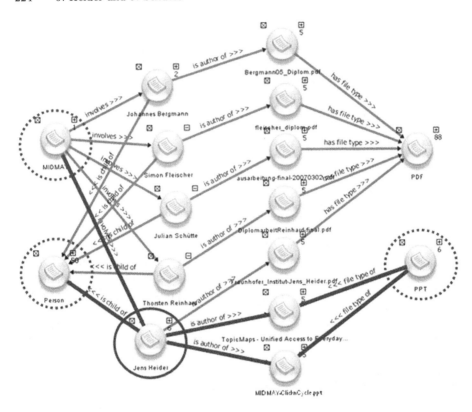

Fig. 5. Calculated result of multiple marked topics in search mode 0

5   Conclusion

Connecting existing data though a unified representation can be used to offer an intuitive way to search for information. The combined interface for browsing, managing and searching, resulting from this concept, is a first step to extend the possibilities users can gain from their stored data. Together with the circular interaction process, the user interface gets independent from current work tasks inside the new information environment. This simplifies the usage and is aligned with the concept of knowledge represented in topic maps.

The concept of marking topics to specify queries allows using the graph theory for searching, thereby leveraging the inherent structure across data sources, created by data occurring in multiple sources. The proof of concept implementation already showed the capabilities created by this approach. Technically, its usage in the field depends on efficient Topic Maps engines that are capable of handling maps containing hundreds of thousands of topics. Compared to the indexing technologies of full-text desktop search engines, this requires currently about ten times more disk space. But by investing these resources, users are not forced to recall specific attributes or keywords and benefit from relations between their stored data. Instead of manually tagging information, the users do not have to invest any effort, but gain the possibility to search directly via keywords, as well as via their own ways of correlating things.

These results are only a first step, but further search techniques with applied graph theory may even improve the approach in terms of flexibility and comprehensibility of the search process. Together with intuitive user interfaces that provide suitable interaction with the user's personal knowledge, the Topic Maps paradigm shows new ways to tackle daily work with information.

References

1. Heider, J.: Vision und Realisierung einer sicheren mobilen Informations-Verteilung, Verwaltung und Abfrage. In: Multikonferenz Wirtschaftsinformatik (MKWI) 2004. Bd 3. Mobile Business Systems (2004)
2. Heider, J., Bergmann, J.: Topicmaps: Unified access to everyday data. In: Proceedings of I-KNOW 2006, Graz, Austria, pp. 473–480 (2006)
3. Wang, D., Dicheva, D., Dichev, C., Akouala, J.: Retrieving information in topic maps: the case of tm4l. In: ACM-SE 45: Proceedings of the 45th annual southeast regional conference, pp. 88–93. ACM Press, New York, NY, USA (2007)
4. Topicmaps.org: XML Topic Maps (XTM) 1.0 Specification (2001), http://www.topicmaps.org/xtm/1.0
5. Barta, R., Salzer, G.: The Tau Model, Formalizing Topic Maps. In: Hartmann, S., Stumptner, M. (eds.) Second Asia-Pacific Conference on Conceptual Modelling (APCCM 2005), Newcastle, Australia. CRPIT, vol. 43, pp. 37–42. ACS (2005), http://crpit.com/confpapers/CRPITV43Barta.pdf
6. Ma, Q., Tanaka, K.: Topic-structure-based complementary information retrieval and its application (2005)
7. Dave, P., Karadkar, U.P., Furuta, R., Francisco-Revilla, L., Shipman, F., Dash, S., Dalal, Z.: Browsing intricately interconnected paths. In: HYPERTEXT 2003: Conference on Hypertext and hypermedia, pp. 95–103. ACM Press, New York (2003)
8. Le Grand, B., Soto, M.: Visualisation of the Semantic Web: Topic Maps visualisation. In: IV 2002: 6th International Conference on Information Visualisation, pp. 344–349 (2002)
9. Ahmed, K.: Beyond PSIs: Topic map design patterns. In: Extreme Markup Languages (2003)

KAIFIA: Knowledge Assisted Intelligent Framework for Information Access

Atta Badii, Chattun Lallah, Oleksandr Kolomiyets, Meng Zhu, and Michael Crouch

IMSS, Intelligent Media Systems and Services Research Centre,
School of Systems Engineering, University of Reading
{atta.badii,c.lallah,o.kolomiyets,meng.zhu,
m.w.crouch}@reading.ac.uk

Abstract. Accessing information, which is spread across multiple sources, in a structured and connected way, is a general problem for enterprises. A unified structure for knowledge representation is urgently needed to enable integration of heterogeneous information resources. Topic Maps seem to be a solution for this problem. The Topic Map technology enables connecting information, through concepts and relationships, and their occurrences across multiple systems. In this paper, we address this problem by describing a framework built on topic maps, to support the current need of knowledge management. New approaches for information integration, intelligent search and topic map exploration are introduced within this framework.

Keywords: Knowledge Representation, Knowledge Discovery, Information Retrieval, Topic Maps, Natural Language Processing.

1 Introduction

A huge amount of information becomes available every day. This information is represented in various formats, usually without any metadata related to it. In such situation the IT community faces the problem of an inefficient information retrieval, which in turn stimulates developing new methods for data mining, semantic acquisition and data presentation across multiple data formats.

Topic Maps is a Semantic Web technology that provides a human-oriented mechanism for information representation and interchanging about the structure of information resources and can be used to solve current problems in knowledge representation. At present Topic Map applications are mainly oriented towards information representation for content management systems, web portals, intranet applications and information integration for enterprises. The scientific interest in Topic Maps is characterised as a simplified approach for knowledge representation in contrast to RDF and OWL for Semantic Web initiatives.

To our knowledge, current solutions for information disclosure are topologically similar to Topic Maps. However, those solutions like Autonomy [4] require additional tools for scalability, interoperability, data merging, etc., which are built-in functionalities of Topic Maps. Different domains, like Health Service, Crime Investigation and Banking, require well-structured, interpreted and connected

L. Maicher and L.M. Garshol (Eds.): TMRA 2007, LNAI 4999, pp. 226–236, 2008.

information to estimate risks, discover hidden dependencies and carry out SWOT[1] analysis.

In this work, we propose a framework that uses Topic Maps as an enabling technology for knowledge representation. This article is organised as follows: Section 2 presents related work on knowledge-based applications and section 3 briefly portrays the Topic Map Technology. Section 4 discusses the future applications of Topic Maps. The main part is presented in Section 5 with the presentation of the proposed Knowledge Assisted Intelligent Framework for Information Access, KAIFIA. Crime Scene Investigation is used as the case study in the description of the framework. Finally Section 6 concludes this paper.

2   Related Work

This section covers the related work on Knowledge Based Applications employing state-of-the-art technologies, including Topic Maps, Ontologies, NLP tools and other semantic frameworks.

Intelligent Topic Manager (ITM). [3] is a software engineering platform for knowledge management, developed by Mondeca[2]. ITM integrates a semantic portal providing the following key functions: Edition, Search, Navigation and Publication. Insight Discover Extractor, IDE, is another tool, which is created by Temis[3] and serves for linguistic analysis and implements a finite-state transducer method that relies on a pre-treatment involving such NLP methods, like document segmentation, lemmatisation and morpho-syntactic analysis of document segments. The semantic portal, built on these two products, employs the information extraction tool and specific linguistic resources provided by Temis and makes use of the domain ontology from Mondeca to populate the knowledge base of the portal in a semi-automatic way. The integration between the information extraction tool and ontological concepts is done by defining the knowledge acquisition rules for mapping words, concepts and their occurrences. This solution is implemented for the domain of legal publishing when the author of legal articles must be informed of every day legal text and court decisions.

HOLMES2. [20] is an investigation management system to assist law enforcement organisations in their management of the complex process of investigating serious crimes. It enables them to improve effectiveness and productivity in crime investigations, helping to solve crimes more quickly and improve detection rates. HOLMES2 is the successor to the HOLMES system (Home Office Large Major Enquiry System), which was a very effective administrative support system for investigating major crimes including serial murders, multi-million pound fraud cases and major disasters.

In a typical major investigation, many documents in the form of statements are produced and all of them have to be carefully processed to ensure that vital clues are

---

[1] Strengths, Weaknesses, Opportunities and Threats.
[2] http://www.mondeca.com/
[3] http://www.temis.com

not overlooked. The system HOLMES2 provides a unique facility that presents information graphically alongside the original documents. Using such a tool users can easily identify important information and link key items together very quickly. In the process of crime investigation many other documents will generate further tasks and HOLMES2 enables the prioritisation of document management and associated actions, manages the allocation and progress of tasks assigned to officers. Through integration with i2[4] and Autonomy software [4], the system provides complex searching tools such as provision to enable automatic notification when particular information is entered and the graphical representation of sequence of events or information within a document.

The office of Naval Intelligence. Everyday, analysts from the Office of Naval Intelligence [12] (ONI) pore through thousand of separate pieces of data, from intercepted phone calls to cryptic messages on web sites. Intelligence analysts must piece that information together in a timely manner to provide effective intelligence for better quality support for operational decision making. Semantic Integration, here, is a task compounded by data streams that may contain different spellings for the same references, terms and expressions, different expressions (surface forms) with the same meaning (deep forms), various meanings for the same words/expressions (synonyms) and associations. For instance, the Library of Congress has detected 32 different ways to spell the name of Muammar Khaddafi. Ontopia[5] provides a Topic Map Based Solution to filter multiple threads of data to identify potentially alarming trends and create a unified and organized analysis. This solution helps ONI organise fragmentary bits and pieces of miscellaneous information semantically to find multiple associations among sets of data, which cannot be done using basic knowledge management tools. It supports ONI analysts to retrieve the precise information they are seeking, and eventually prepare more timely, detailed and effectual reports.

3 Topic Maps Technology

Topic Maps (TM) are organised around *topics*, and each topic is used to represent a real-world object. A detailed description of Topic Maps is available in [2, 10, 17]. Our proposed framework is based on the use of Topic Maps for adding semantic structure to the data collected from unstructured data sources of information. The framework also incorporates the architecting of networks comprising competing and collaborating communities of intelligent Crawlers and Foragers, configured as eco-systems, whose survival would depend on the quality of information gathered in terms of appropriate relevance and quality criteria formulated in terms of an evaluation function [6, 5]. The Topic Map Technology best suits the authoring of statements and information resulting from the crime scene investigation domain. In this domain, loads of information is collected from multiple heterogeneous sources. This information is of multiple modalities, may be in the form of textual/audio or video statements from victims and suspects, video from CCTV cameras, and other endless information

---

[4] i2 offers an integrative suite of visual investigative analysis. See http://www.i2inc.com
[5] http://www.ontopia.net/index.html

sources. Maintaining and managing this amount of information has been very challenging for the intelligence policing unit. HOLMES2 provides a unique interface to annotate crimes by linking information extracted from different information resources. However, HOLMES2 does not link the information based on its semantic concepts, but relies solely on user decisions and data interpretation. The application of Topic Maps here would confer much value in semantic integration and knowledge linking, thus providing an automated and therefore inexpensive way of connecting disconnected information, which may be fragmented across various sources with complex form variations and on a large scale.

Topic Maps provide means of structuring the semantic concepts of variably sourced and variously formed root expressions relating to police observations and/or witness statements and evidence integration in crime-investigation statements. The concepts extracted from the statements can be easily linked through semantic reasoning to other cases, thus enabling knowledge integration, open sense making and linking both within and across the collaborating agencies and communities of practice involved. Data between departments can be exchanged in a unified XML format and each crime investigation unit will have a Unified Data View of a crime scene. Such a Unified View will enable the Investigator to discover information from multiple perspectives, as all concepts found in the topic maps will be related based on linguistic semantic relationships, user perspectives, domain-specific ontologies, added common-sense reasoning to extract missing links between concepts, as described in KAIFIA.

4 Future Applications of Topic Maps

Topic Map is currently used in many information systems environments, spanning a spectrum of applications such as E-Government, E-Learning, E-Commerce, Crime Tracking, and many other applications, as listed in [16]. Information Today[6] has released the results of a study regarding Enterprise Search [19]: "The study found that 59% of respondents plan on upgrading or enhancing their search solutions in the relatively near future. The most desired additions, named by survey respondents were *Automatic Categorisation* (34%), followed by *Topic Maps* (31%), *Results Clustering* (30%), and *Visualisation of Results* (29%)." According to this study, the advent of Topic Maps heralds an emerging technology for future Knowledge Management Applications.

Until now, the main utility of Topic Maps can be summarised in the following terms: Content Classification; Semantic Search; Information Sharing [11]. However, the recent research and development on the Topic Map Technology is yet to establish a dominant framework, which can be used for future applications to meet new challenges like:

- Automatic Topic Map Population
- Reasoning in Topic Maps for intelligent interactive applications

---

[6] Information Today, Inc. (ITI) is a leading publisher and conference organizer in the information and knowledge management industries, http://www.infotoday.com/

- Visualising complex and large "infospheres" using advanced HMI[7]
- Intelligent searches based on ad hoc notions of concept "similarity" and "interestingness" from various users' standpoints, different viewpoints therein, and, under variable contexts.

Our proposed Knowledge Assisted Intelligent Framework for Information Access, called KAIFIA, aims at addressing these challenges in a generic manner so as to radically empower future knowledge-based applications.

5 Knowledge Assisted Intelligent Framework for Information Access, KAIFIA

KAIFIA, as illustrated in [Fig. 1] consists of the following layers: Data Source, Knowledge, Intelligent Reasoning, System Interface and the Hardware Layer. The framework provides a global solution, exemplified in the context of Crime Investigation, as a demonstrator test bed for KAIFIA, as explained in this section.

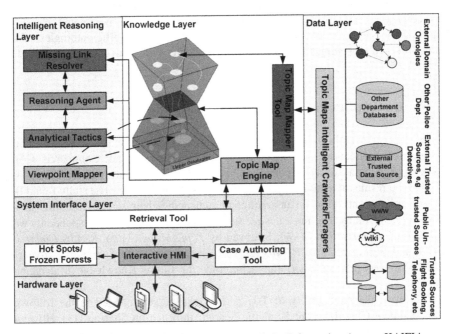

Fig. 1. Knowledge Assisted Intelligent Framework for Information Access, KAIFIA

5.1 The KAIFIA Data Source Layer

The KAIFIA *Data Source Layer* consists of a number of information channels, which provide data of different formats. Such channels may be topic maps from internal

---

[7] Human Machine Interfaces.

and/or external trusted sources, enterprises, publicly available information on the Web, such as Wikis, Newsgroups etc., and other trusted sources, like Flight Booking databases and Telephony logs. External Domain Ontologies constructed by domain experts also serve as an input to the knowledge layer of the framework.

Topic Map Intelligent Crawler/Foragers: Crawlers were first introduced as a kind of automatic information gathering tool, comprising software programs that can browse the WWW in a methodical, automated manner [13]. Foragers and Crawlers are different on the basis of their search goals in terms of whether these were simply based on a generalization ontology defining some notion of interestingness of a source or data given to the agent (i.e. a Forager) to define a search mission in unbounded spaces (internets and intranets) , or alternatively specific terms given to an agent (i.e. a Crawler) to search specific distributed databases in open spaces (an internet or intranet) to find any relevant information [5]. The web-crawlers have been successfully implemented as a central part of search engines for facilitating gathering information on the web and subsequent indexing of the collected Web pages. Since this technique has been successfully applied for more than a decade in commercial applications like, RBSE [8], Google Crawler [7], WebCrawler [18], etc, we therefore propose to introduce crawlers into enterprise-level Topic Map search in order to automate the time and labour consuming tasks of knowledge acquisition.

The proposed intelligent TM crawler starts from a hotspot, i.e. a topic of interest within the topic map, to mine the topic map constantly conforming to certain searching criteria based on the user's request. We believe that such a crawler over topic maps provides a more intelligent way of searching as it can traverse nested structures deep within the topic map, thus enables a robust and informed search through multiple levels of abstraction. For example, in the context of criminal data intelligence gathering and linking, the challenge posed by our demonstrator test bed, employing this autonomous TM crawler facilitates the forensic investigation in terms of evidence gathering, clue spotting, hotspots monitoring etc, as the TM crawler is not limited to the topic map space only, but can also crawl deeper into the data resource layer via the occurrences of topics, e.g. finger print matching and face identification. Such a solution is extremely beneficial for cross-police-force or cross-country serious organised crime investigation.

5.2 The KAIFIA Knowledge Layer

The KAIFIA *knowledge layer* exploits the use of the Topic Map Technology to create a rich knowledge base, which consists of Upper Ontologies, Domain Specific and Domain Generic Ontologies.

Topic Map Mapper Tool: The information gathered by the TM Intelligent Crawler, does not conform to a standard format as described in the Data Source Layer. The Topic Map Mapper tool is responsible for unlocking the value of the data located in disparate locations and merging them into one central knowledge base. It also assigns an initial trust level value, based on the trustworthiness of the data source; such as, if the data is coming from an internal data source, its value of trust would be higher than the value of trust assigned to data coming from the public web.

Topic Map Engine: The framework we are proposing incorporates a robust Topic Map Engine, which interfaces the core system to the users. The Engine enables the processing of huge topic maps and provides interfaces for optimised querying and update of the Topic Map Repository. There are many available tools and applications, which can be extended for this purpose in the framework. These tools include the TM4J Engine[8], the Ontopia Topic Map Engine[9], and many others, as listed in [21].

5.3 The KAIFIA Intelligent Reasoning Layer

The KAIFIA reasoning layer is the vital part of the framework that confers the ability for intelligent search and reasoning over the semantic integration of the search results. Without reasoning, the KAIFIA knowledge base, would serve as a data repository with static logics, like conventional databases. The *Intelligent Reasoning Layer* provides automated tools for semantics enrichment of the data and consists of the following tools:

Missing Links Resolver: Employing upper and domain-specific ontologies, the *Missing Link Resolver* helps to identify potential links between topics within a case. In this context, an approach analogous to linguistic semantics frames [9] or rules from upper ontologies [15] can be deployed, e.g. a rule for the "buying" action contains an equal action of "selling", which is semantically the same event, but interpreted from another viewpoint. The Missing Links Resolver works together with the *Reasoning Agent*. The Reasoning Agent composes the case from the contextual data and suggests similar significant cross-cases commonalities to the user. The Resolver also supports background queries, known as *monitoring*, posted by the user, which shape the criteria for the required information, and as soon as such information is registered within the framework, the user will be notified and then can complete the case annotation.

Analytical Tactics: *Analytical tactics* use domain-specific scenarios to conceptualise case-based patterns for intelligent navigation through knowledge bases. For example, the UK Police Service approach to the investigation of serial and serious crimes is based on a standard process [20]. According to this process our framework provides an interface that allows users to conceptualise the data entered into the knowledge base by making use of domain and task-specific scripts. After the scripts run, the framework can suggest options, e.g. pre and post-conditions, which need to be fulfilled in order to proceed to the next step in the investigation.

Viewpoint Mapper: Data from the information sources can be represented in different formats and distributed across an organisation. The information presented to the user needs to be dependent upon the individual user's purpose and current viewpoint. The individual users may wish to construct and customise their own personal knowledge space. The *Viewpoint Mapper* serves as a unified data presentation as well as personalised user viewpoints, according to the user role, nature

---

[8] The TM4J Topic Map Engine, http://tm4j.org/tm4j-engine.html
[9] The Ontopia Topic Map Engine, http://www.ontopia.net/solutions/engine.html

of information etc. by making use of conceptual scaling as defined in [14]. The different users' viewpoints (perspectives) are described in section 5.4.2.

5.4 The KAIFIA System Interface Layer

Retrieval Tool: The KAIFIA *Retrieval Tool* serves to create, edit and post queries to the KAIFIA knowledge base via the Topic Map Engine. The retrieval tool can support Free-text, Suggest, Advanced, Fuzzy, and Structured queries. The results of querying can be transferred to the Interactive HMI Tool for advanced interaction with users or presented in one of the following ways: hyperlinked-trees, graphs, landscapes or worlds [1].

Interactive HMI: Additionally, using the Topic Map Technology as the back-end for the indexing of resources through concepts allows navigation in a Virtual Knowledge Space, enabled by an *Augmented HMI*. This Virtual Knowledge Space is a universe of semantic concepts, through which the user can navigate to discover new knowledge. Such environment introduces new ways of knowledge disclosure. These include *knowledge exploration, chance discoveries,* and *discovering chances.* "Chance Discovery" is a term we use to distinguish the fact that when a user is navigating through a known or unknown region in the TM network, he/she may discover relevant information which he/she was not expecting to find at that place or at that time. On the other hand, "Discovering Chances" refers to the user's experiential knowledge of relative interestingness of sources/infospheres in finding relevant information per their past experience and personal knowledge of such sources/infospheres. Having previous knowledge of a region in the topic map, the user may know in advance the best way of finding the information he/she is looking for. In other words, the user is learning from experience by continuing to discover new chances of finding the relevant information as they are exposed to the outcomes of each search effort. Besides that, the Interactive HMI will employ the state-of-the-art technologies of NLP and Dialog Management to add value to the intelligent interactivity in searching and browsing for new knowledge.

Case Authoring Tool: Authoring TM has also been a problem hindering the widespread adoption of this technology in the information editorial world. This is because user-friendly interactive tools are not available to ease the population of TM. Through the *Case Authoring Tool*, the Interactive HMI (previously described) can be used to enable easy and interactive ways of describing crime statements and adding other crime related information to the knowledge base.

5.4.1 KAIFIA Semiotic Contextualisation
Semiotic Contextualisation can be leveraged in a user-intuitive fashion to allow attention economy and directed focus by the user in navigating topic maps and particularly in traversing TM nodes in a user-specified sequence when exploring large topic maps. This can be achieved by adopting a colour coding, such as a traffic light colour scheme to denote any arbitrary distinction relevant to the way the user would wish to explore the topic map, e.g. can designate *Hot Spots* or *Frozen Forests*.

KAIFIA Hot Spots: *Hot Spots* is a new technique, we are proposing, to solve the problem of navigating and searching through complex and massive topic maps. The hot spots in the topic maps are defined by attributes, like user defined notions of "interestingness", search contexts and common-sense reasoning provided by the framework. Fig. 2 shows the colour washes applied to a TM on the basis of some arbitrary contextual information. The user will be able to selectively explore the different regions, each time traversing through and viewing only on the basis of certain criteria of interestingness of the moment.

KAIFIA Frozen Forests: These give the user the ability to segregate any part of the search space in a nested fashion and designate them as frozen forests that the user may wish to leave well alone. Paradoxically, frozen forests can be designated anywhere even deep inside hot spots, thus allowing a re-focusing of attention recursively inside any region to accelerate the search and exploration. Other colours or

Fig. 2. Colour Washes for Topic Map Hot Spots Visualisation

semiotic contextualisation could be used together with the above to denote other aspects such as "strangeness", "older"/"younger" data, etc. In this way, single nodes or whole areas can be given a particular colour wash (contextualisation to expedite exploration). This facility greatly enhances the ability of the user to define perspectives in a variety of ways and to exploit radical empowered selective search, visualisation and knowledge integration and linking within and across various levels, viewpoints and abstraction spaces.

5.4.2 KAIFIA Multiple Perspectives

The target interface space can deploy the power of semiotic encoding as exemplified above by using the colour codes to allow dynamic user definable perspectives (colour washes) over any number of nodes or clustering of nodes to suit each user's fleeting and idiosyncratic exploration goals of the moment. In this way such "colour washes" can designate any number of topic maps anywhere in the search space as belonging to any one of the following perspectives:

Contextual Perspective: This is a perspective whereby a number of nodes are related because they share the same designated semiotic contextualisation, which may be ascribed to any node or constituencies of nodes by the user for arbitrary reasons e.g. share the same property of "interestingness", "similarity" from a particular viewpoint in user's worldhood. Folding graph techniques can be deployed to bring together physically dispersed nodes on the same screen for a viewing as may be exercised by the user, whilst signalling the actual distance of such nodes from each other if they happen to be widely dispersed.

Processual Perspective: This is a perspective whereby nodes with some logical and causal relationship are annotated using the spectral range to denote temporal sequence and/or spatio-temporal abstraction or subsumption (low level or high level fusion in time or space to allow causal linkages over these attributes to be semantically integrated and reasoned over).

Semantic Perspective: This is a perspective whereby any grouping of proximal nodes are arbitrarily assigned a certain semantic label to serve a user's on-the-fly perceptual management of the exploration space while searching a TM with a particular search goal. When such viewpoints become activated they can allow the on-the-fly and ad hoc meta-labelling of nodes or node clusters to aid the perceptual management of the evolving search space as the user continues to search for a priority topic, whilst attempting to manage opportunistic search by responding to serendipity.

5.5 The KAIFIA Hardware Layer

The *Hardware Layer* in the framework enables communication between users and the knowledge space by supporting all major protocols with a direct access to the retrieval tool through the *Interactive HMI*. One of the main stakeholders is the community of editors, populating the Knowledge base with new information, like new case statement, new crime scene descriptions, new audiovisual descriptions and other related information. The second stakeholder accessing this layer in the framework is the community of users, who need well-structured and rich information, such as crime investigators and police officers as in our demonstrator example application domain. These users can access the knowledge-base through different digital devices, like PDAs, Smart Phones, etc.

6 Conclusion

In this paper, an enabling Knowledge Integration tool environment has been introduced as an enhanced knowledge accessibility, visualisation and open sense making tool, KAIFIA. This is presented as a solution for intelligent and knowledge-based applications, which require a high level of accelerated information integration, interpretation and presentation. The emphasis of the framework is aimed at providing an infrastructure for Topic Map Centred Knowledge Bases, along with intelligent reasoning agents, and interactive HMI. The solution poses new challenges and in response to these, we have proposed new techniques for knowledge engineering, using Topic Maps. These techniques highlight the Intelligent TM Crawler, Missing links Resolver, Viewpoint Mapper and Analytical Tactics. New TM exploration techniques, like Hot Spots and Frozen Forests, should empower the user by providing a variety of user-configurable and user definable dynamic perceptual space management tools to meet the user-cognitive and computation and visualisation challenges of high-speed exploration and dynamic re-focusing management of search spaces whilst traversing through multiple linked repositories.

References

1. Ahmed, K.: Topic Maps for Repositories (2000),
   http://www.gca.org/papers/xmleurope2000/papers/s29-04.html
2. Ahmed, K.: An Introduction to Topic Maps (April 2007),
   http://www.networkedplanet.com/technology/topicmaps/intro.html
3. Amardeilh, F., Laublet, P., Minel, J.: Document annotation and ontology population from linguistic extractions. In: Proc. 3rd ACM Int. Conf. on Knowledge Capture, pp. 161–168 (2005)
4. Autonomy Software (Accessed, April 2007), http://www.autonomy.com/content/home/
5. Badii, A.: Integrated Data Discovery, Exchange Engineering and Mining Management Architecture using communities of foraging agents. In: Proc. 1st Int. Conf. on Knowledge Transfer (1996)
6. Badii, A.: Architecting agent ecologies for collaborative-competitive-creative communities: Facilitating opportunistic learning, improvisation, innovation and diffusion. In: Proc. 3rd European Conf. on Knowledge Management (2002)
7. Brin, S., Page, L.: The anatomy of a large-scale hypertextual Web search engine. Computer Networks and ISDN Systems 30(1-7), 107–117 (1998)
8. Eichmann, D.: The RBSE spider: balancing effective search against Web load. In: Proc. 1st WWW Conf. (1994)
9. Fillmore, C.J.: Frame semantics and the nature of language. In: Conf. on the Origin and Development of Language and Speech, vol. 280, pp. 20–32 (1976)
10. Garshol, L.: What are Topic Maps (2002),
    http://xml.com/pub/a/2002/09/11/topicmaps.html?page=1
11. Garshol, L., Moore, G.: Topic Maps Workshop, Everything you ever wanted to know (2006), http://forum.dataforeningen.no/attachment.php?attachmentid=619
12. Isogen, I.: Organizing Mountains of Information into Actionable Data: ONI relies on Topic Maps to improve search capabilities (Accessed, April 2007), http://innodata-isogen.com/resources/case_studies/oni_cs.pdf
13. Kobayashi, M., Takeda, K.: Information retrieval on the web. ACM Computing Surveys 32(2), 144–173 (2000)
14. Neuss, C., Kent, E.R.: Conceptual Analysis of Resource Meta-information. In: Proc. 3rd Int. WWW conf. on technology, tools, and applications (1994)
15. Niles, I., Pease, A.: Towards a Standard Upper Ontology. In: Proc. 2nd Int. Conf. on Formal Ontology in Information Systems, pp. 17–19 (2001)
16. Ontopia, Applications of Topic Maps,
    http://www.ontopia.net/solutions/oks-applications.html
17. Pepper, S.: The TAO of Topic Maps: Finding the way in the age of infoglut (Accessed, April 2007), http://www.ontopia.net/topicmaps/materials/tao.html
18. Pinkerton, B.: Finding what people want: Experiences with the WebCrawler. In: Proc. 1st WW W Conf. (1994)
19. Search Engine Applications Focus of New Research Report by Information Today (2007), http://findarticles.com/p/articles/mi_m0EIN/is_2007_March_29/ai_n18766900
20. What is HOLMES2 (Accessed, April 2007), http://www.holmes2.com/holmes2/whatish2/
21. Woodman, M.: Topic Map Tools, http://www.topicmap.com/topicmap/tools.html

Report from the Open Space and Poster Sessions*

Lars Marius Garshol[1] and Lutz Maicher[2]

[1] Bouvet ASA, Oslo, Norway
larsga@bouvet.no
http://www.bouvet.no
[2] University of Leipzig, Johannisgasse 26, 04103 Leipzig
maicher@informatik.uni-leipzig.de
http://www.informatik.uni-leipzig.de/~maicher/

Abstract. This is a summary of the presentations made in the poster session and the two open space sessions at the TMRA 2007 conference. The poster session consists of peer-reviewed conference submissions accepted as posters, while the open space sessions were free-form sessions where anyone could sign up to do a short presentation. The result is a collection of reports on works in progress and interesting ideas, some of which are likely to appear as papers at next year's conference.

1 The Poster Session

This report summarizes the five contributions to the poster session of TMRA 2007, mainly based on the submitted abstracts, but also on the actual posters, and impressions from the presentations. All posters were submitted to the Program Committee, refereed and accepted as poster presentations for the conference. All posters were exhibited during the whole conference and presented in a series of ten minute presentations during the poster session. (Besides the officially accepted posters described here some ad-hoc posters were put up by the conference attendees.) The poster session was moderated by Rani Pinchuk. All posters are available at the conference website.

1.1 Automatic Topic Maps Generation from Free Text Using Linguistic Templates

Ann Houston reports[1] on the formulation and use of linguistic templates to identify and extract Topic Maps components (topics, occurrences, associations) from free English text that represents a coherent discourse domain, e.g., travel guides. A topic map is automatically created using these extractions, and the

---

* To appear 2007 in: Maicher, Lutz & Garshol, Lars Marius (eds.): *Scaling Topic Maps*. Proceedings of TMRA'07 - International Workshop on Topic Map Research and Applications: Leipzig, Germany, October 11-12, 2007; LNCS 4999, Springer-Verlag, see http://www.springeronline.com/lncs

[1] http://www.informatik.uni-leipzig.de/~tmra/2007/slides/houston_TMRA2007.pdf

L. Maicher and L.M. Garshol (Eds.): TMRA 2007, LNAI 4999, pp. 237–251, 2008.

resulting map is evaluated for utility and accuracy. The linguistic templates are refined and extended, based on the initial topic map. This approach hopes to provide a rapid and efficient way of boot-strapping useful topic maps through an iterative cycle of free text extraction, topic map generation and template enhancement. It should be mentioned, that due to the use of linguistic patterns, the presented approach is not-knowledge free, which means that it bundled on a specific language, and to some extent, on a specific domain.

1.2 Towards Holistic Knowledge Creation and Interchange Part II: Examples, Theory and Strategy

Dino Karabeg and Roy Lachica argue that meta-information development, and more generally creation and organization of knowledge and insights within and across communities and disciplines, will not be the result of simple automation of current practices, but of systematic redesign involving technology, theory and practice. WiKeyPoDia.org here serves both as prototype example and as symbol of this approach, showing how familiar Web 2.0 techniques can be applied to grow information in a new, vertical direction, namely towards structure and insight rather than volume. A theoretical basis that supports this approach is outlined, as well as examples of application in several areas including education, academic communication and democratic decision-making. Combined with the accompanying talk [Karabeg08], the results allow the authors to make progress towards the projected Holoscopia platform for holistic knowledge creation and interchange.

1.3 Why Aren't Topic Maps Ruling the World Yet?

Tobias Redmann, Hendrik Thomas and Bernd Markscheffel note[2] that half a decade has passed since Eric Freese asked "So why aren't Topic Maps ruling the world?" [Freese02]. (It was observed several times at the conference, that the correct form of the question is: "So why *isn't* Topic Maps ruling the world?". The poster's authors argue that their headline cites the phrase by Eric Freese from 2002.)

Much progress has been made, e.g. the finalization of the ISO Topic Map standardization process. However, obviously Topic Maps still do not rule the world. Asking for reasons, they note that many problems Freese mentioned are still insufficiently solved and new problems have emerged, preventing Topic Maps from unleashing its full potential. Many things had been promised and prototypically implemented but the breakthrough in using and accepting Topic Maps as a powerful knowledge managing concept in science and business is still missing.

The poster (due to the provided interactive components, like notes, pencils and post-its, was also referred to as a "poster 2.0" by the conference attendees) points out these problems, in order to start a new fruitful discussion process in the

---

[2] http://www.informatik.uni-leipzig.de/~tmra/2007/slides/redmann1_
TMRA2007.pdf

Topic Maps community. The diverse discussion facilities at the poster are divided into the sections development, use-cases and applications, modelling questions, information and marketing, basic research and standards, and miscellaneous (because everything is miscellaneous).

1.4 Ontology-Driven Strategy for Inference of Learning Object Metadata

Reidar Bratsberg, Jan Schreiber, and Terje Syversen argue[3] that considering a system with a large number of learning objects represented in a topic map, the work involved in markup of detailed metadata is prohibitively complex and time-consuming. In their poster they discuss a method for placing metadata and inference strategies into the topic map to allow reuse based on structures defined in the ontology. Their use case is GREP, the Norwegian national curriculum represented as a topic map. The learning resources of this curriculum span 200 different e-learning web sites and over 200,000 topics in three different topic maps with different ontologies.

1.5 Design Principles for a Topic Maps Wiki - The Wiki Way of Knowledge Management with Topic Maps

Tobias Redmann and Hendrik Thomas underline[4] that wikis have proven themselves as a fast and easy approach to knowledge management. In the poster they discuss similarities between, or better complements of, Topic Maps and the wiki concept. Based on their conclusions they present sparse design principles for a Topic Maps-based Wiki. The objective is to provide a "wiki way" of Topic Maps creation in order to enhance semantic knowledge management. Although with Topincs [Cerny08] such a system already exists (and was used for semantic, collaborative knowledge management at the TMRA 2006[5] and 2007[6] conferences), a bunch of such approaches were widely discussed in the Open Space sessions.

2   The Open Space Sessions

The following is a summary of the open space sessions in chronological order based on the slides used by the presenters, a blog summary of the conference ([Garshol07a] and [Garshol07b]), impressions from the conference itself, and email discussions with the presenters.

The contributions were informal and non-refereed, since workshop attendees had been given the opportunity to sign up to short talks on a flip chart during the conference, and the suggested format for each presentation was five minutes

---

[3] http://www.informatik.uni-leipzig.de/~tmra/2007/slides/bratsberg_TMRA2007.pdf

[4] http://www.informatik.uni-leipzig.de/~tmra/2007/slides/redmann_TMRA2007.pdf

[5] http://www.topincs.com/tmra/2006/

[6] http://www.topincs.com/tmra/2007/

presentation, and five minutes discussion. Both sessions were moderated by Lars Marius Garshol. The outcome of this "playground for visionaries" is this report on forward-looking work in progress.

2.1   Topics as Document Proxies

Peter-Paul Kruijsen presented an approach to content management used internally by the Dutch consulting company Morpheus. Morpheus stores content in the version control system Subversion. Subversion allows a normal file system structure of files and directories to be versioned in a central repository, much like CVS, SCCS and other version control systems.

The content in Morpheus's Subversion repository is not just source code, but all internal content, which includes things like presentations, meeting minutes, etc. Morpheus has added a hook to Subversion, using the standard Subversion extension feature for this, which is run every time a file is added to the repository. The hook is written in Java using the OKS API and an open source Subversion API, and adds a topic representing the file to a topic map which describes the repository.

The hook uses a Subversion ontology developed by Morpheus to described files in the repository, as in this example LTM fragment:

```
#PREFIX svn @"http://psi.mssm.nl/svn/"
[doc123 : svn:file = "/meetings/20070607-pk-google.txt"
  %"https://svn.mssm.nl/morpheus/meetings/20070607-pk-google.txt"]
svn:authorship(doc123 : svn:file, pk : svn:author)
{pk, svn:authorname, [[pk]]}
{doc123, svn:date-added, [[2007-06-21 13:42:24 (revision 8)]]}
{doc123, svn:date-modified, [[2007-06-27 09:23:06 (revision 10)]]}
```

The topic map is then maintained manually using TM-ADMIN, Morpheus's Topic Maps editor, which is a Java web application based on Ontopia's OKS engine. In TM-ADMIN the user adds another type to Subversion files from a domain-specific ontology. The topic type for the file in the example above might be "meeting minutes". The TM-ADMIN editor is driven by a schema which specifies constraints for different types of files, so that once the "meeting minutes" type has been added the schema requires information about where the meeting was, when, and with whom. The user must then fill in this metadata, and if new topics have to be created the schema requires certain minimal information about these as well, and so on.

The topic map can be used for information retrieval by browsing it in a simple Omnigator-like browsing application. Certain workflow support features have been implemented, like a query that retrieves all instances of `svn:file` that are not instances of any other type. These are all files which need to be annotated by the user, and this gives users an easy way to see what needs to be done.

2.2   Development of TM4J 2.0

Xuân Baldauf presented his work on TM4J 2.0, an open source Topic Maps engine written in Java. TM4J is the main open source Topic Maps engine, and the 2.0 version is a rewrite of the core of the engine to add full support for the new ISO standards such as TMDM and XTM 2.0. The API has been updated to match the TMDM wording and now uses generics.

Implementation classes in the engine have 5 layers of interfaces and super-classes. For example, the topic implementation class is put together as follows:

- `ReadableTopic`: A basic topic interface, containing only accessor methods, and no mutator methods.
- `Topic`: An extension of `ReadableTopic` that adds the missing mutator methods.
- `BasicTopic`: An implementation of `Topic` that contains the data for a single topic instance.
- `MergedTopic`: A read-only implementation of `ReadableTopic` that contains the union of the data for a set of merged `BasicTopics`.
- `TopicImpl`: A wrapper class adding compatibility with the TM4J 0.9 API, allowing older code to continue to work against TM4J 2.0.

TM4J 2.0 contains two implementations of the API that appear identical to the developer, but which have different performance and persistence characteristics. The simplest is a straightforward in-memory implementation that simply keeps all data in instance objects and has no built-in support for persistence. Persistence can of course be supported by means of import/export to and from Topic Maps formats like XTM. The other implementation extends the in-memory implementation with persistence support, via the db4o object database [db4o] over JPOX. JPOX is a JDO implementation, for persistent storage of the topic maps [JPOX].

At the time of the open space sessions the TM4J 2.0 engine was not completely finished, but close.

2.3   Linking Topic Maps Engines

Graham Moore and Marc Wilhem Küster presented their thoughts on the importance of being able to exchange Topic Maps data between Topic Maps engines. This can of course be done by means of simple XTM export and import, but this does not in itself support more advanced scenarios like synchronization, where one engine receives the changes made to a part of a topic map maintained in another engine. In practice, synchronization is a much more important scenario than simple export and import.

NetworkedPlanet has developed a solution to this where a Topic Maps engine can publish an Atom feed [Atom] describing a sequence of modifications made to the topic map. Each item in the feed describes a single transaction, that is, a set of related modifications to the topic map. Each transaction is given a GUID, and if the changes are transmitted into another feed by the client engine the

GUID should be preserved in order for downstream engines to determine if they see the same transaction more than once. This mechanism is being used in the context of the nascent European eGovernment Resource Network ([Kuster08]) to synchronize between nodes.

Their work on fragment interchange has highlighted an issue with both XTM 1.0 and 2.0: any fragment is going to contain fully described topics, ontology topics, and stub topics. The difficulty is that there is no defined way to communicate in the fragment which topics belong to which category. This is important because although the stub topics will have statements made about them, they are not fully described, and will need to be loaded from the remote source in order to become fully described. Seeing whether topics have statements is not a sufficient test. The best solution to this issue appears to be to communicate the identities of fully described topics outside the XTM fragment itself, but this is of course less clean.

2.4   TMRM in OWL

Jack Park gave a presentation on his work on representing the TMRM in OWL. He found he couldn't do it OWL DL, but it was possible in OWL Full, which was a blessing in disguise, since OWL Full can do so much more. The reason given for switching to OWL Full is that the TMRM requires that each property type used to represent a subject must, itself, be a subject in the map. OWL separates properties from classes, and the TMRM requires that properties also be classes. This complicates applications written in OWL; they can validate to OWL Full, but not to OWL DL.

He didn't show slides, but showed RDF/XML in a tool called TopicSpaces-OWL, a Java application that he's written himself. The map includes subjects that support Tagomizer [Park07]. The application, a work in progress, includes an incomplete editor, and a TouchGraph-based graph visualizer [TouchGraph]. It uses Jena for RDF manipulations [Jena], the H2 database for storage [H2], Lucene for full-text indexing [Lucene], and is primarily an exploration tool for developing topic maps in OWL using the the SubjectProxy/SubjectProperty representations of one variant of the TMRM specification.

2.5   High-Performance Topic Maps

Axel Borge described an approach to caching of Topic Maps query results developed by Graham Moore and himself as part of the project to develop PIA, the Topic Maps-based intranet for the Norwegian Mail. The intranet has 26,000 users, each one of which has a highly personalized home page. This means that the home pages cannot be effectively cached, and given the amount of expected traffic this would require the Topic Maps engine to perform 800 queries per second. Since the topic map contains roughly 2 million TAOs this would not be feasible without some form of caching.

The solution they came up with is called CacheQube, and is really a way to cache query results where cached results are dropped when changes are made to the topic map that invalidate the results. The difficulty here is to work out exactly which changes those are, and the CacheQube does this by tracking all (topic, association type) pairs that are examined during the execution of a query. When updates are made this information is used to determine whether or not to drop a particular cached query result.

The TMCore Topic Maps engine uses SQL as its query language, based on a set of predefined views and functions called TMRQL [Park07]. Queries are therefore executed by the database, and so TMCore cannot itself track the (topic, association type) pairs. The developer must therefore define this set for each query in order for the CacheQube to work.

Still, the approach was successful in the Norwegian Mail project, and it is now being incorporated as part of the TMCore product.

2.6   Semantic Search with Topic Maps

Lars Marius Garshol presented an approach to "semantic search" using Topic Maps. Most Topic Maps implementations today support full-text search, but this is only full-text search as used by Google and databases, and while results are improved by the Topic Maps structure it does not really make full use of the semantics of Topic Maps. He showed a prototype of a system that attempts to really use the semantics to interpret natural language queries from users.

The approach taken in the prototype is to parse a query against the names of topics in the topic map, and to use the ontology of the topic map to construct possible interpretations of the query. The interpretations are expressed in tolog [Garshol06a] and executed to produce the search result.

A demonstration was given, using a topic map of personal photos. It was showed how a search for "photos of sam oh" would return exactly that. Internally, the query is parsed into a set of two topics: photo (a topic type) and Sam Oh (an instance of the person type). The system assumes this means the user is searching for instances of the type related to the given instance, based on an analysis of the ontology, and thus produces the correct answer.

The system understands subtyping, so that queries for, say, a place, will find all kinds of places, like countries, cities, etc. The system also understands hierarchies (given correct ontology annotation), so that queries for "photos of sam oh in canada" will work, even though the photos will be tagged with individual locations in Canada, and not have any direct associations with Canada.

The prototype was developed in Jython using the OKS. It's not publicly accessible anywhere, because more work is needed to make it robust enough to withstand use by normal users.

2.7   Scoping Subject Identifiers

Lutz Maicher and Xuân Baldauf spoke on contextual subject identity, expanding on Maicher's earlier work on the subject in [Maicher06]. They argued that subject identity is dependent on perspective, while the TMDM merges topics based on subject identifiers unconditionally. They pointed out that there could be reasonable disagreement on how many subjects a set of topics corresponded to, and that scoping subject identity would allow the set of topics to be contextually dependent.

This sparked intense discussion at the conference, and also in a conference blog. Reactions were nearly universally negative, and the arguments were generally that subject identity is too fundamental to Topic Maps for this to be allowed, and that identity is never contextual.

2.8   Topic Maps and Web 3.0

Graham Moore spoke on his vision of what Web 3.0 might look like. He described it as a semantic web with improved global information access based on Topic Maps and bottom up collaboration. It would build on Web 2.0's easy authoring and minimal user interfaces, and make use of lessons learned over the past 10 years in a clean and simple fashion.

He gave some impressions of what he thought Web 3.0 would be like:

- It would be RESTful.
- Topics would be addressable, and have many representations such as XTM, RDF/XML, and plain XML. They would also have ATOM feeds, as described in section 2.3 on page 241.
- Ontologies would be used to provide semantic search, dynamic user interfaces, and semantic agreement. These features would be self-configured based on the ontology and user context.
- Links would be two-way and typed, as in Topic Maps.
- Controlled vocabularies would be shared.
- Topic Maps would replace folksonomies and tagging. This implies the creation of topic clouds to replace tag clouds.
- Existing web CMSs would be replaced by Topic Maps-based systems with better metaphors, better authoring processes, and much better search. Graham called this Knowledge-Centric Publishing (KCP).

2.9   Topic Maps-Based Wiki

Lars Marius Garshol presented a design for a Topic Maps-based wiki, and showed a prototype of it. The basis of the design is the observation that wikinames and topics are fundamentally the same, in that both represent a single subject, and are used as a place to gather all information about that subject. However, wikis lack structure, and Topic Maps lack content. Combining structure and content solves that and yields a very powerful knowledge management tool.

The editor is Ontopoly, with the wiki content being written in the description occurrence field, using wiki markup. The wiki markup has been extended to support links to topics and inline tolog queries, which are displayed as tables. The queries are crucial, as they allow structured overviews of content, exploiting the Topic Maps structure. The wiki view looks much like a normal wiki, with a central text area and a "fact box" a la Wikipedia. The fact box, however, is generated from the topic map's associations and occurrences.

This simple approach has been successfully used by the presenter for a topic map on OKS feature requests and ISO standards issues.

2.10   Knowledge Is a Mountain

Dino Karabeg argued that knowledge is structured like a mountain, rising out of what he called "the information jungle". He claims knowledge is not a simple set of concepts and relations, but that some concepts and relations are more important than others, and therefore need to be more visible than others, corresponding to the top of the mountain. What people need, is knowledge that is structured like the mountain, in order to avoid drowning in a mass of undifferentiated data. Dino sees this as heralding a paradigm shift, and cites many sources to substantiate the claim.

Dino phrased his title as an equation:

$$knowledge = mountain \qquad (1)$$

From this, he drew two corrolaries, the first being:

$$knowledge \neq jungle \qquad (2)$$

According to Dino, this has consequences for Topic Maps. It is important not to build topic maps that simply interconnect things in the jungle, as that will only make the jungle thicker.

The second corrolary was:

$$killer\ app = mountain\text{-}building\ toolkit \qquad (3)$$

In other words, if Topic Maps are to realize their full potential they need to scale the mountain.

2.11   A Style Language for Topic Maps

Hendrik Thomas made a call for a style language for visualization of Topic Maps, claiming that the current graph-based visualizations were insufficient. Instead, different visualizations for different kinds of topic maps are needed. For this a style language is necessary, in which the visualizations can be configured.

Hendrik also showed a prototypical visualization tool, called tmchartis, wherein visualizations could be set up manually, based on the topic map structure.

2.12   What's Wrong with Merging

Rani Pinchuk gave a presentation on what he perceives as a weakness of merging: it only adds information, it cannot remove information. Therefore, merging is not sufficient to support synchronization of Topic Maps between different servers. For example, if two users work on separate copies of a topic map and both make different changes, then if they merge their changed topic maps into a single topic map, deleted and changed information will still be presented in the merged topic map.

Rani pointed out that this problem occurs in real life, for example when using TopiWriter [Pinchuk07]. In TopiWriter topic maps are stored within Word files, and these are often edited independently by different users, making it difficult to reconcile the copies of the topic map afterwards.

Rani further pointed to proposed solutions like TMSync [Garshol07c] and the use of ATOM feeds (section 2.3 on page 241), and argued that these were not standardized, which means that they are not truly interoperable.

2.13   PSIs for Versioning

Xuân Baldauf presented a Topic Maps annotation vocabulary he called "PSIs for Versioning", describing it as a solution to the problem raised in section 2.12 on the preceding page. Xuân's example use case was an address book, which typically gets updated independently on several different devices before synchronization. Again, the problem is the same as described by Rani: merging simply makes deleted or changed data come back.

Xuân's proposed solution is to annotate all statements in the topic map, using reification, as shown below in a version of CTM:

```
xuan
   phonenumber: "+41 76 2169639" ~ asrt0 .
valid(what: asrt0, state: entered, since: date20070619)
valid(what: asrt0, state: deleted, since: date20070922)
```

This states that the occurrence was entered on June 19th, and deleted on September 22nd. Applications which understand and respect the annotation ontology will avoid the synchronization problem, and at the same time gain support for versioning.

In the discussion it was suggested to simplify this to simply scoping deleted statements with a past topic. This has the benefit of being more likely to be compatible with other applications.

Like Rani, Xuân observed that there was no standardized solution for this problem, and wondered if this solution would be suitable for standardization.

2.14   COBOL and Topic Maps

Dmitry Bogachev presented his ideas for what a Topic Maps-based programming language might look like, inspired by COBOL. The presentation was later expanded on in a blog posting [Bogachev07]. COBOL had built-in support for defining and manipulating persisted data, something later programming languages for the most part have eschewed, effectively outsourcing it to databases.

Object-oriented languages remedy this to some degree, allowing information about entities to be modelled directly, and, if object-oriented databases are used, to be persisted. This allows code to be written as in the following Ruby example:

```
class Person
  attr_accessor :first_name, :last_name
end

p = Person.new
p.first_name = 'John'
p.last_name = 'Smith'
```

This, however, does not allow statements to be annotated, to cater for data changing over time, coming from different sources, and so on. Nor is there any direct support for inferencing or access control. This can of course be supported, but in order to do so the developer has to build a substantial infrastructure and integrate it, and the client API is complicated considerably by this.

With dynamic languages it's possible to use meta-programming to at least lessen the impact on the client API, allowing code like this:

```
a=Person.create(:psi=>'JohnSmith',:at_date => '2007-10-01')
a.first_name='John'
a.last_name='Smith'
a.save

b=Person.create(:psi=>'JoeSmith',
                        :first_name=>'Joe',
                        :last_name=>'Smith',
                        :at_date => '2007-10-02')
b.save
a=Person.find(:psi=>'JohnSmith',:at_date => '2007-10-03')
b=Person.find(:psi=>'JoeSmith',:at_date => '2007-10-03')
b.knows_person << a
a.save
b.save
```

However, said Dmitry, a Topic Maps-based programming language could do even better, given that support for this kind of annotation and inferencing would be part of its core.

2.15 A Graphical Notation for Topic Maps

Maik Pressler presented his beginning thesis work on this, and emphasized that he was aware of the existing GTM proposal from ISO [ISO13250-7]. However, he felt that it was too oriented towards traditional data modelling, and very similar to UML and ER. He would have liked to see something simpler, and more comparable to mind-mapping.

He listed a set of personal requirements for GTM that largely intersect with those defined by ISO:

- Must be easy to understand, also for those who are not Topic Maps experts.
- Must be suitable for sketches. (Also an ISO requirement.)

- Must be suitable for printing on paper. (Also an ISO requirement.)
- Must be usable for discussing, explaining, and developing Topic Maps. (Also an ISO requirement.)
- Must support sharing Topic Maps.

In his thesis he plans to define requirements for graphical notations for Topic Maps, evaluate existing proposals, and make his own proposal.

2.16   Reification Versus Annotation

Lutz Maicher brought up an issue with reification, which is that there are two possible ways to reify a statement. One is to create a topic representing the relationship represented by the statement, as shown in the example below in LTM syntax:

```
employed-by(lm : employee , ul : employer) ~ employment-lm
{employment-lm , startdate , [[2002-11-01]]}
```

The second line is clearly a statement about `lm`'s employment, and not about the association itself, and Lutz called this reification. However, other possible approach is to talk about the association itself, which Lutz called *annotation*. This is shown in the example below (using the same syntax as reification for now):

```
employed-by(lm : employee , ul : employer) ~ employment-lm
created-by(employment-lm : creation, user-x : creator)
```

In this example the reifying topic is clearly representing the *statement*, and not the employment. However, the TMDM very clearly states that reification is always to be interpreted as in the former example above.

Lutz argued that this leaves reification as defined in the TMDM unable to support a very common use case, which is annotation of statements (as, for example, done in section 2.13 on page 246), and proposed a convention for the second form of reification. In this convention statement reification is done by making a subject locator of the reifying topic be the same as an item identifier of the statement to be reified. LTM has no syntatic construct for expressing this directly, but the fragment below shows how it can still be done:

```
employed-by(lm : employee , ul : employer)
[employment-lm %"#employed-by"]
created-by(employment-lm : creation, user-x : creator)
```

In the discussion there was universal consensus that this was the right approach.

2.17   Topic Maps Wiki

Benjamin Bock gave another presentation on the subject of Topic Maps and wikis. His proposal is different from than in section 2.9 on page 244 in that

Benjamin proposed writing Topic Maps fragments in CTM directly into the wiki markup, rather than using a Topic Maps editor to edit a topic map containing wiki markup. An example might look as follows:

```
Benjamin_Bock
This is a Article about
[-fullname:"Benjamin Bock"].
[-firstname:"Benjamin"] is the author of
[authoring(author: self,software: RTM) "RTM"].
He was born in [yob:1982].
```

The TM fragments are extracted from the wiki source and concatenated using one newline (n) character. A special identifier, self, refers to the topic which is described by this wiki page. The last part of a statement is the display name which may be enhanced with knowledge obtained from the fragment or an ontology. The result would be the following CTM:

```
Benjamin_Bock
-fullname:"Benjamin Bock"
-firstname:"Benjamin"
yob:1982

.

authoring(author: Benjamin_Bock,software: RTM)
```

Rendered, this would look as follows, using icons to show the types of assertions:

Benjamin Bock

This is a Article about ᴀᴀ<u>Benjamin Bock</u>.
ᴀᴀ<u>Benjamin</u> is the author of ⊛<u>RTM</u>.
He was born in ⊕<u>1982</u>.

Fig. 1. Screenshot

The Topic Maps fragments would be parsed, creating a topic map. New statements would be unscoped, while statements in old versions would be scoped as old. The set of statements in an entire wiki would effectively make up a single topic map.

This approach could be integrated into tools like MediaWiki and used, for example, to create a mash-up of Wikipedia.

References

[Garshol07a]   Garshol, L.M.: TMRA – day 1; blog entry, Larsblog (October 11, 2007), http://www.garshol.priv.no/blog/132.html

[Garshol07b]   Garshol, L.M.: TMRA – day 2; blog entry, Larsblog (October 13, 2007), http://www.garshol.priv.no/blog/133.html

[Garshol07c]   Garshol, L.M.: Synchronizing Topic Maps with External Sources. In: Maicher, L., Sigel, A., Garshol, L.M. (eds.) TMRA 2006. LNCS (LNAI), vol. 4438, pp. 192–199. Springer, Heidelberg (2007)

[Garshol06a]   Garshol, L.M.: Tolog – A Topic Maps query language. In: Maicher, L., Park, J. (eds.) TMRA 2005. LNCS (LNAI), vol. 3873, pp. 183–196. Springer, Heidelberg (2006)

[Kuster08]   Küster, M.W., Moore, G.: Scaling Topic Maps. In: Maicher, L., Garshol, L.M. (eds.) TMRA 2007. LNCS (LNAI), vol. 4999, pp. 1–13. Springer, Heidelberg (2008)

[Maicher06]   Böhm, K., Maicher, L.: Real-Time Generation of Topic Maps from Speech Streams. In: Maicher, L., Park, J. (eds.) TMRA 2005. LNCS (LNAI), vol. 3873, pp. 112–124. Springer, Heidelberg (2006)

[Atom]   Nottingham, M., Sayre, R.: RFC 4287 - The Atom Syndication Format. Internet Engineering Task Force (December 2005), http://tools.ietf.org/html/rfc4287

[Park07]   Park, J.: Tagomizer: Subject Maps Meet Social Bookmarking. In: Maicher, L., Park, J. (eds.) TMRA 2005. LNCS (LNAI), vol. 3873. Springer, Heidelberg (2006)

[Park07]   Moore, G., Ahmed, K.: Topic Map Relational Query Language - TM-RQL NetworkedPlanet Ltd. (2005), http://www.networkedplanet.com/download/TMRQL.pdf

[Pinchuk07]   Pinchuk, R., Aked, R., de Orus, J.-J., De Weerdt, D., Focant, G., Fontaine, B., Wolff, M.: TopiWriter - Integrating Topic Maps with Word Processor. In: Maicher, L., Sigel, A., Garshol, L.M. (eds.) TMRA 2006. LNCS (LNAI), vol. 4438, pp. 184–191. Springer, Heidelberg (2007)

[Bogachev07]   Bogachev, D.: Subject-centric programming language or what was good about COBOL, Subject-centric blog (October 23, 2007), http://www.subjectcentric.com/post/Subject-centric_programming_language_or_what_was_good_about_COBOL

[ISO13250-7]   ISO/IEC WD: 13250-7: Topic Maps – Graphical Notation for Topic Maps, International Organization for Standardization, Geneva, Switzerland (July 4, 2007), http://www.jtc1sc34.org/repository/0882.pdf

[db4o]   db4objects inc., db4objects, http://www.db4o.com/

[JPOX]   Java Persistent Objects project. JPOX, http://www.jpox.org/

[Lucene]   Apache Software Foundation, Lucene, http://lucene.apache.org/

[H2]   HSQLDB Group, H2 Database Engine, http://www.h2database.com/html/main.html

[TouchGraph]   TouchGraph project, TouchGraph, http://sourceforge.net/projects/touchgraph/

[Jena]   Jena project, Jena Semantic Web Framework, http://jena.sourceforge.net/

[Cerny08]   Cerny, R.: Topincs Wiki. A Topic Maps Powered Wiki. In: Maicher, L., Garshol, L.M. (eds.) TMRA 2007. LNCS (LNAI), vol. 4999, pp. 57–65. Springer, Heidelberg (2008)

[Karabeg08]   Lachica, R., Karabeg, D.: Metadata Creation in Socio-semantic Tagging Systems: Towards Holistic Knowledge Creation and Interchange. In: Maicher, L., Garshol, L.M. (eds.) TMRA 2007. LNCS (LNAI), vol. 4999, pp. 162–171. Springer, Heidelberg (2008)

[Freese02]   Freese, E.: So why aren't Topic Maps ruling the world? In: Proceedings of Extreme Markup, IDEAlliance, Montréal, Canada (2002), http://www.idealliance.org/papers/extreme/proceedings/html/2002/Freese01/EML2002Freese01.html

Author Index

Badii, Atta   226
Barnickel, Thorsten   116
Barta, Robert   98
Bhatti, Nadeem   51
Bock, Benjamin   172
Bode, Stephan   36
Brecht, Rike   36

Cahier, Jean-Pierre   154
Cerny, Robert   57
Crouch, Michael   226

de Azevedo, Renato Preigschadt   86
Drobnik, Oswald   128

Garshol, Lars Marius   25, 74, 237
Godehardt, Eicke   51
Guittard, Claude   154

Heider, Jens   214
Henriques, Pedro Rangel   86
Hofmann, Tobias   66

Johnsen, Lars   41

Karabeg, Dino   160
Kolomiyets, Oleksandr   226
Küster, Mare Wilhelm   1

Lachica, Roy   160
Lallah, Chattun   226
Librelotto, Giovani Rubert   86

Maicher, Lutz   198, 237
Markscheffel, Bernd   36
Moore, Graham   1

Nenova, Karamfilka   116

Park, Jack   140
Pepper, Steve   186
Pradella, Martin   66

Ramalho, José Carlos   86

Schütte, Julian   214
Smolnik, Stefan   14
Spekowius, Karsten   36
Stümpflen, Volker   116

Thomas, Hendrik   36

Ueberall, Markus   128

Zaher, L'Hédi   154
Zhu, Meng   226

Lecture Notes in Artificial Intelligence (LNAI)

Vol. 5113: P. Eklund, O. Haemmerlé (Eds.), Conceptual Structures: Knowledge Visualization and Reasoning. X, 311 pages. 2008.

Vol. 5110: W. Hodges, R. de Queiroz (Eds.), Logic, Language, Information and Computation. VIII, 313 pages. 2008.

Vol. 5108: P. Perner, O. Salvetti (Eds.), Advances in Mass Data Analysis of Images and Signals in Medicine, Biotechnology, Chemistry and Food Industry. X, 173 pages. 2008.

Vol. 5097: L. Rutkowski, R. Tadeusiewicz, L.A. Zadeh, J.M. Zurada (Eds.), Artificial Intelligence and Soft Computing – ICAISC 2008. XVI, 1269 pages. 2008.

Vol. 5078: E. André, L. Dybkjær, W. Minker, H. Neumann, R. Pieraccini, M. Weber (Eds.), Perception in Multimodal Dialogue Systems. X, 311 pages. 2008.

Vol. 5077: P. Perner (Ed.), Advances in Data Mining. XI, 428 pages. 2008.

Vol. 5076: R. van der Meyden, L. van der Torre (Eds.), Deontic Logic in Computer Science. X, 279 pages. 2008.

Vol. 5064: L. Prevost, S. Marinai, F. Schwenker (Eds.), Artificial Neural Networks in Pattern Recognition. IX, 318 pages. 2008.

Vol. 5040: M. Asada, J.C.T. Hallam, J.-A. Meyer, J. Tani (Eds.), From Animals to Animats 10. XIII, 530 pages. 2008.

Vol. 5032: S. Bergler (Ed.), Advances in Artificial Intelligence. XI, 382 pages. 2008.

Vol. 5027: N.T. Nguyen, L. Borzemski, A. Grzech, M. Ali (Eds.), New Frontiers in Applied Artificial Intelligence. XVIII, 879 pages. 2008.

Vol. 5012: T. Washio, E. Suzuki, K.M. Ting, A. Inokuchi (Eds.), Advances in Knowledge Discovery and Data Mining. XXIV, 1102 pages. 2008.

Vol. 5009: G. Wang, T. Li, J.W. Grzymala-Busse, D. Miao, A. Skowron, Y. Yao (Eds.), Rough Sets and Knowledge Technology. XVIII, 765 pages. 2008.

Vol. 5001: U. Visser, F. Ribeiro, T. Ohashi, F. Dellaert (Eds.), RoboCup 2007: Robot Soccer World Cup XI. XIV, 566 pages. 2008.

Vol. 4999: L. Maicher, L.M. Garshol (Eds.), Scaling Topic Maps. XI, 253 pages. 2008.

Vol. 4994: A. An, S. Matwin, Z.W. Raś, D. Ślęzak (Eds.), Foundations of Intelligent Systems. XVII, 653 pages. 2008.

Vol. 4953: N.T. Nguyen, G.S. Jo, R.J. Howlett, L.C. Jain (Eds.), Agent and Multi-Agent Systems: Technologies and Applications. XX, 909 pages. 2008.

Vol. 4946: I. Rahwan, S. Parsons, C. Reed (Eds.), Argumentation in Multi-Agent Systems. X, 235 pages. 2008.

Vol. 4944: Z.W. Raś, S. Tsumoto, D.A. Zighed (Eds.), Mining Complex Data. X, 265 pages. 2008.

Vol. 4938: T. Tokunaga, A. Ortega (Eds.), Large-Scale Knowledge Resources. IX, 367 pages. 2008.

Vol. 4933: R. Medina, S. Obiedkov (Eds.), Formal Concept Analysis. XII, 325 pages. 2008.

Vol. 4930: I. Wachsmuth, G. Knoblich (Eds.), Modeling Communication with Robots and Virtual Humans. X, 337 pages. 2008.

Vol. 4929: M. Helmert, Understanding Planning Tasks. XIV, 270 pages. 2008.

Vol. 4924: D. Riaño (Ed.), Knowledge Management for Health Care Procedures. X, 161 pages. 2008.

Vol. 4923: S.B. Yahia, E.M. Nguifo, R. Belohlavek (Eds.), Concept Lattices and Their Applications. XII, 283 pages. 2008.

Vol. 4914: K. Satoh, A. Inokuchi, K. Nagao, T. Kawamura (Eds.), New Frontiers in Artificial Intelligence. X, 404 pages. 2008.

Vol. 4911: L. De Raedt, P. Frasconi, K. Kersting, S. Muggleton (Eds.), Probabilistic Inductive Logic Programming. VIII, 341 pages. 2008.

Vol. 4908: M. Dastani, A. El Fallah Seghrouchni, A. Ricci, M. Winikoff (Eds.), Programming Multi-Agent Systems. XII, 267 pages. 2008.

Vol. 4898: M. Kolp, B. Henderson-Sellers, H. Mouratidis, A. Garcia, A.K. Ghose, P. Bresciani (Eds.), Agent-Oriented Information Systems IV. X, 292 pages. 2008.

Vol. 4897: M. Baldoni, T.C. Son, M.B. van Riemsdijk, M. Winikoff (Eds.), Declarative Agent Languages and Technologies V. X, 245 pages. 2008.

Vol. 4894: H. Blockeel, J. Ramon, J. Shavlik, P. Tadepalli (Eds.), Inductive Logic Programming. XI, 307 pages. 2008.

Vol. 4885: M. Chetouani, A. Hussain, B. Gas, M. Milgram, J.-L. Zarader (Eds.), Advances in Nonlinear Speech Processing. XI, 284 pages. 2007.

Vol. 4874: J. Neves, M.F. Santos, J.M. Machado (Eds.), Progress in Artificial Intelligence. XVIII, 704 pages. 2007.

Vol. 4870: J.S. Sichman, J. Padget, S. Ossowski, P. Noriega (Eds.), Coordination, Organizations, Institutions, and Norms in Agent Systems III. XII, 331 pages. 2008.

Vol. 4869: F. Botana, T. Recio (Eds.), Automated Deduction in Geometry. X, 213 pages. 2007.

Vol. 4865: K. Tuyls, A. Nowe, Z. Guessoum, D. Kudenko (Eds.), Adaptive Agents and Multi-Agent Systems III. VIII, 255 pages. 2008.

Vol. 4850: M. Lungarella, F. Iida, J.C. Bongard, R. Pfeifer (Eds.), 50 Years of Artificial Intelligence. X, 399 pages. 2007.

Vol. 4845: N. Zhong, J. Liu, Y. Yao, J. Wu, S. Lu, K. Li (Eds.), Web Intelligence Meets Brain Informatics. XI, 516 pages. 2007.

Vol. 4840: L. Paletta, E. Rome (Eds.), Attention in Cognitive Systems. XI, 497 pages. 2007.

Vol. 4830: M.A. Orgun, J. Thornton (Eds.), AI 2007: Advances in Artificial Intelligence. XIX, 841 pages. 2007.

Vol. 4828: M. Randall, H.A. Abbass, J. Wiles (Eds.), Progress in Artificial Life. XII, 402 pages. 2007.

Vol. 4827: A. Gelbukh, Á.F. Kuri Morales (Eds.), MICAI 2007: Advances in Artificial Intelligence. XXIV, 1234 pages. 2007.

Vol. 4826: P. Perner, O. Salvetti (Eds.), Advances in Mass Data Analysis of Signals and Images in Medicine, Biotechnology and Chemistry. X, 183 pages. 2007.

Vol. 4819: T. Washio, Z.-H. Zhou, J.Z. Huang, X. Hu, J. Li, C. Xie, J. He, D. Zou, K.-C. Li, M.M. Freire (Eds.), Emerging Technologies in Knowledge Discovery and Data Mining. XIV, 675 pages. 2007.

Vol. 4811: O. Nasraoui, M. Spiliopoulou, J. Srivastava, B. Mobasher, B. Masand (Eds.), Advances in Web Mining and Web Usage Analysis. XII, 247 pages. 2007.

Vol. 4798: Z. Zhang, J.H. Siekmann (Eds.), Knowledge Science, Engineering and Management. XVI, 669 pages. 2007.

Vol. 4795: F. Schilder, G. Katz, J. Pustejovsky (Eds.), Annotating, Extracting and Reasoning about Time and Events. VII, 141 pages. 2007.

Vol. 4790: N. Dershowitz, A. Voronkov (Eds.), Logic for Programming, Artificial Intelligence, and Reasoning. XIII, 562 pages. 2007.

Vol. 4788: D. Borrajo, L. Castillo, J.M. Corchado (Eds.), Current Topics in Artificial Intelligence. XI, 280 pages. 2007.

Vol. 4775: A. Esposito, M. Faundez-Zanuy, E. Keller, M. Marinaro (Eds.), Verbal and Nonverbal Communication Behaviours. XII, 325 pages. 2007.

Vol. 4772: H. Prade, V.S. Subrahmanian (Eds.), Scalable Uncertainty Management. X, 277 pages. 2007.

Vol. 4766: N. Maudet, S. Parsons, I. Rahwan (Eds.), Argumentation in Multi-Agent Systems. XII, 211 pages. 2007.

Vol. 4760: E. Rome, J. Hertzberg, G. Dorffner (Eds.), Towards Affordance-Based Robot Control. IX, 211 pages. 2008.

Vol. 4755: V. Corruble, M. Takeda, E. Suzuki (Eds.), Discovery Science. XI, 298 pages. 2007.

Vol. 4754: M. Hutter, R.A. Servedio, E. Takimoto (Eds.), Algorithmic Learning Theory. XI, 403 pages. 2007.

Vol. 4737: B. Berendt, A. Hotho, D. Mladenic, G. Semeraro (Eds.), From Web to Social Web: Discovering and Deploying User and Content Profiles. XI, 161 pages. 2007.

Vol. 4733: R. Basili, M.T. Pazienza (Eds.), AI*IA 2007: Artificial Intelligence and Human-Oriented Computing. XVII, 858 pages. 2007.

Vol. 4724: K. Mellouli (Ed.), Symbolic and Quantitative Approaches to Reasoning with Uncertainty. XV, 914 pages. 2007.

Vol. 4722: C. Pelachaud, J.-C. Martin, E. André, G. Chollet, K. Karpouzis, D. Pelé (Eds.), Intelligent Virtual Agents. XV, 425 pages. 2007.

Vol. 4720: B. Konev, F. Wolter (Eds.), Frontiers of Combining Systems. X, 283 pages. 2007.

Vol. 4702: J.N. Kok, J. Koronacki, R. Lopez de Mantaras, S. Matwin, D. Mladenič, A. Skowron (Eds.), Knowledge Discovery in Databases: PKDD 2007. XXIV, 640 pages. 2007.

Vol. 4701: J.N. Kok, J. Koronacki, R. Lopez de Mantaras, S. Matwin, D. Mladenič, A. Skowron (Eds.), Machine Learning: ECML 2007. XXII, 809 pages. 2007.

Vol. 4696: H.-D. Burkhard, G. Lindemann, R. Verbrugge, L.Z. Varga (Eds.), Multi-Agent Systems and Applications V. XIII, 350 pages. 2007.

Vol. 4694: B. Apolloni, R.J. Howlett, L. Jain (Eds.), Knowledge-Based Intelligent Information and Engineering Systems, Part III. XXIX, 1126 pages. 2007.

Vol. 4693: B. Apolloni, R.J. Howlett, L. Jain (Eds.), Knowledge-Based Intelligent Information and Engineering Systems, Part II. XXXII, 1380 pages. 2007.

Vol. 4692: B. Apolloni, R.J. Howlett, L. Jain (Eds.), Knowledge-Based Intelligent Information and Engineering Systems, Part I. LV, 882 pages. 2007.

Vol. 4687: P. Petta, J.P. Müller, M. Klusch, M. Georgeff (Eds.), Multiagent System Technologies. X, 207 pages. 2007.

Vol. 4682: D.-S. Huang, L. Heutte, M. Loog (Eds.), Advanced Intelligent Computing Theories and Applications. XXVII, 1373 pages. 2007.

Vol. 4676: M. Klusch, K.V. Hindriks, M.P. Papazoglou, L. Sterling (Eds.), Cooperative Information Agents XI. XI, 361 pages. 2007.

Vol. 4667: J. Hertzberg, M. Beetz, R. Englert (Eds.), KI 2007: Advances in Artificial Intelligence. IX, 516 pages. 2007.

Vol. 4660: S. Džeroski, L. Todorovski (Eds.), Computational Discovery of Scientific Knowledge. X, 327 pages. 2007.

Vol. 4659: V. Mařík, V. Vyatkin, A.W. Colombo (Eds.), Holonic and Multi-Agent Systems for Manufacturing. VIII, 456 pages. 2007.

Vol. 4651: F. Azevedo, P. Barahona, F. Fages, F. Rossi (Eds.), Recent Advances in Constraints. VIII, 185 pages. 2007.

Vol. 4648: F. Almeida e Costa, L.M. Rocha, E. Costa, I. Harvey, A. Coutinho (Eds.), Advances in Artificial Life. XVIII, 1215 pages. 2007.

Vol. 4635: B. Kokinov, D.C. Richardson, T.R. Roth-Berghofer, L. Vieu (Eds.), Modeling and Using Context. XIV, 574 pages. 2007.